OPEN SCIENCE:

Distance Teaching and Open Learning
of Science Subjects

OPEN SCIENCE:

DISTANCE TEACHING AND OPEN LEARNING OF SCIENCE SUBJECTS

by
SHELAGH ROSS AND EILEEN SCANLON

P·C·P
Paul Chapman
Publishing Ltd

All rights reserved
Paul Chapman Publishing Ltd
144 Liverpool Road
London
N1 1LA

British Library Cataloguing in Publication Data

A B C D E F G H   9 8 7 6 5

Ross, Shelagh
Open Science: Distance Teaching and Open
Learning of Science Subjects
I. Title II. Scanlon, Eileen
507.1041

ISBN 1–85396–172–8

Typeset by Palimpsest Book Production Limited,
Polmont, Stirlingshire
Printed and bound in Great Britain by
Athenaeum Press Ltd, Gateshead, Tyne & Wear.

CONTENTS

PREFACE

Distance education is a growth industry worldwide. Open learning and improved access to higher education are high on the political agendas of many countries. The recruitment success of the distance education institutions has raised awareness of the large number of mature students keen to enter further and higher education, and the last two decades have seen the development of distance teaching and open learning to extents that have fostered totally new attitudes to the provision of educational opportunities at these levels. An upsurge of interest in the design of instructional materials is now resulting in multiple media techniques, many of which originated in the distance education sector, being increasingly applied in mainstream schools, colleges and universities, as well as in industrial and vocational training.

Yet in some quarters there is still a feeling that remote learning is feasible only for certain types of knowledge-based subject, and unsuitable for areas such as science that require the development of practical skills (or indeed for areas that focus heavily on the development of interpersonal skills). There are also those who doubt the contention that a strongly hierarchical subject like science can be taught at degree level to students accepted on an 'open entry' policy, i.e. without entrance qualifications. Our aim in this book is to refute the notion that science is a discipline that is impossible (or at best very difficult) to teach at a distance and/or more or less from scratch at university level.

Many of the examples we quote involve material or data produced at the British Open University (UKOU). As the first national, autonomous, degree-awarding institution to provide courses solely in the distance teaching mode, the UKOU was initially, and continues to be, highly influential in the field of distance education. Over the past twenty-five years it has generated a great deal of institutional research, and many major contributions to the literature of distance and open education have been based on its systems. The Open University has offered science degrees from its inception, and thousands of science graduates have contributed to its position as the largest single provider of higher education in the UK. Currently it produces one-sixth of the annual output of the country's graduates. In 1993, a number equivalent to 10 per cent of full-time students starting university level science in Britain were beginning to study science with the Open University. The UKOU also happens to be the institution with which we, as science educators, are most familiar. Inevitably therefore, we have drawn heavily on this institution's experiences and practices to illustrate certain points. However, we have attempted to maintain a balance by including references to other systems and institutions, covering a range of educational levels, countries and cultures. Our aim has been to consider the role of distance and open learning techniques not just in the institutions specifically geared to them, but also within the mainstream of higher education. Throughout, we have been concerned to shift the focus away from details of any one 'system' towards the aims of distance educators, the design of instructional materials for distance and open learning in science subjects, the uses of multiple media techniques and the evaluation of the educational effectiveness of individual schemes. Thus we hope that practitioners across the whole spectrum of science education will find ideas and techniques that are relevant to their particular situation.

To set the material in context, Part A contains three background reviews:

- on the theory and practice of distance and open education;
- on the use of remote learning specifically for education in the scientific disciplines;
- on perspectives in educational psychology as applied to the design of instructional materials.

Thereafter, we dwell less on philosophies or policies than on practice. Part B describes many forms of practice, at different levels, showing how distance education techniques can be applied to science teaching. The main emphasis is on the actual design and presentation of instructional materials, and the experience of learners in using such materials. Finally, Part C offers some pointers to the future.

We are greatly indebted to the many Open University students who over the past 15 years or so have shared with us their wonder and excitement at discovering the world of science. We have also been very fortunate in the many colleagues, too numerous to mention individually, with whom we have worked to design and teach a variety of courses and who have helped us to reflect on this experience. We particularly wish to pay tribute to the late Mike Pentz, who did so much to shape the Open University's Science Faculty and who initiated both of us into the mysteries of teaching science at a distance. In writing this book, we have made many references to examples of UKOU course material, and we gratefully acknowledge all those course teams whose work we have quoted. We also wish to express our appreciation of the help given us by the following colleagues who have commented on draft chapters or otherwise contributed significantly to our thinking: Alison Ashby, Phil Butcher, Keith Harry, Barbara Hodgson, Adrian Kirkwood, Patricia Murphy and Tim O'Shea. We thank Dave Perry for his assistance with the production of Figures 3.2 and 12.4, and John Trapp for similar help with Figure 8.3.

ACKNOWLEDGEMENTS

We are grateful to the copyright holders for permission to reproduce material:

Fig 3.2 Springer-Verlag

Figs 4.1, 6.1 Routledge

Fig 5.1 Crown copyright

Figs 5.2, 5.3, 5.4, 5.5, 5.6, 5.7, 5.8, 5.9; 6.2, 6.3, 6.4, 6.5, 6.6, 6.7, 6.8; 7.2, 7.3, 7.4, 7.5, 7.6; 8.3, 8.5; 10.1, 10.2; 11.1, 11.2, 11.3, 11.4, 11.5, 11.6, Table 11.1; 12.6; 14.2 The Open University

Fig 6.9 The Institute of Electrical and Electronics Engineers Inc.

Fig 8.1 McGraw-Hill Inc.

Fig 8.4 Morgan Kaufman Inc.

PART A:

Background

1

DISTANCE EDUCATION AND OPEN LEARNING

Distance education is a method of education ... The concept of open learning describes the nature of the education.

(Rumble, 1989a, p.28)

1. INTRODUCTION

Distance education is not a new concept. The first private correspondence colleges were established in the mid-1800s, mainly to teach students who wanted to sit public examinations, such as those for University of London external degrees. By the end of the last century, there followed accredited correspondence programmes at a few public institutions: for example, the University of Chicago diversified into correspondence courses in 1891. France has had a central, government-funded, distance education provision for 50 years, covering all levels from primary to postgraduate. Both Australia and Canada have an extensive tradition of correspondence teaching for compulsory schooling of pupils living in remote areas. Some such schemes were expanded after the Second World War into 'Schools of the Air', with two-way radios allowing direct interactive communication between teachers and children. Australia has also led the way since the middle 1950s in the establishment of mixed mode university structures serving both campus-based and remote students. Despite this background, however, it was not until the 1970s and 1980s that the growth in distance education mushroomed on an international scale. Today the value of distance education is recognized by governmental policy decisions and funding in many countries. Ideas and techniques developed within the distance teaching sector have permeated educational thinking across the board. In this chapter we will look at the recent evolution of distance and open education systems, and examine some of the foundations on which they are built.

The worldwide upsurge of interest in the theory and practice of distance education has spawned a considerable literature. One of the problems associated with the international nature of this literature is the lack of consistency in the terminology used over the years in different countries and systems. We start therefore, in typical scientific tradition, by 'defining our terms'.

2. DISTANCE EDUCATION

The debate about what exactly does and does not constitute distance education is important, not so much because the naming of parts matters, but because analysis of the various definitions can of itself contribute to a theoretical framework and provide examples of good practice. Such an analysis was published by Keegan in 1980 and updated in 1990 in the light of the responses provoked by the first version. Keegan's final synthesis highlights five essential characteristics of distance education:

1. The 'quasi-permanent' separation of students from their teacher(s) throughout the educative process.
2. The 'quasi-permanent' separation of each learner from other learners on the same programme throughout the educative process, with students receiving individual rather than group-based instruction.

This separation of teachers and learners is a fundamental feature of all definitions of distance education, clearly distinguishing distance teaching from contiguous teaching. Keegan's choice of the term 'quasi-permanent' to denote the isolation of each learner is a recognition of the fact that some face-to-face contacts (meetings between individual learners and tutors or counsellors, self-help gatherings of students, seminars, laboratory classes, field trips, even brief residential schools) are not precluded. The extent of such contacts varies from system to system, on a continuum that ranges from non-existent, through marginal and/or voluntary, to essential.

3. The centrality of an educational organization in planning, producing and distributing the learning materials, and in co-ordinating student support services.

This characteristic separates distance education from non-institutionalized, unsupported, individual study, such as learning from 'teach-yourself' books or cassettes, or by self-directed reading. It also highlights a crucial aspect of distance teaching which Keegan (1983, p. 13) has encapsulated as 'In traditional education, a teacher teaches. In distance education an institution teaches. This is a radical difference.' This distinction has profound implications for any face-to-face sessions that may take place, implications that will be examined at length in Chapter 13. Here we will simply note Perraton's 1983 summary (p. 40) of one consequence: 'Where a tutor meets distance students face-to-face, [his or her] role is changed from being a communicator of information to that of being a facilitator of learning.'

4. The provision of two-way communication that allows the student to participate in dialogue, both as initiator and as respondent.

The inclusion of two-way communication as an essential feature of a distance education system again distinguishes distance education from the 'teach-yourself' programmes.

5. The use of technical media to transmit course content and to link students with teachers.

In this context, print is regarded as a technical medium in the same way as audio or videocassettes, broadcasting, telecommunications, and the various computer-based technologies.

Three elements that appeared in many of the early descriptions of distance education but have not been included in Keegan's latest definition are worthy of further comment. These are the issues of independence, privatization and industrialization.

Definitions put forward in the 1970s and early 1980s often laid stress on ideals such as learner independence and freedom. Nowadays it is recognized that many distance education systems are in fact highly structured and inflexible, leaving little learner autonomy. This is an issue that we will examine more closely in the context of open learning. But if independence in learning has dropped out of the definitions of distance education, the concept of privatization of learning is beginning to creep in. Smith (1987) has pointed out that developed societies increasingly value products that promote private lifestyles – the recent steep rise in the ownership of video-players is just one example of this trend. While totally self-directed study is the most private form of learning, it may not be a feasible route to the particular knowledge, skills or qualification that the learner requires. If these can only be attained through recognized institutions, distance education institutions offer a more private kind of learning than the mainstream systems. This privacy factor, while not part of the formal definition, is nevertheless a characteristic feature of distance education, and one that almost certainly contributes to its current popularity, at least in Western societies.

The concept of distance education as an industrialized form of teaching and learning was first fully set out by Peters in 1973. A detailed article appeared in English translation in 1983. Peters' interpretation has attracted a good deal of attention and controversy, and continues to be extensively discussed in the literature. Part of the industrialization process is fairly obvious in the sense that the production phase of the operation involves long-range planning, financial investment, division of labour (between, for example, academics, media specialists, editors and printers), and the application of organizational principles to the production process. Depending on the size of the institution, it will benefit to a greater or lesser degree from the economies of scale that come from mass-production. The management skills required in distance education are also more akin to those needed in industrial organizations than to those normally exercised in conventional educational establishments. Even these kinds of statement have been met with dissent in some quarters, but in a review of his position in 1989 Peters has been at pains to point out that his concept extends, perhaps more controversially still, into the teaching and learning processes. His thesis is that within the complex organizational structure of a distance education system, tutors and counsellors are just one part of the whole – important but nevertheless only one component. In this sense their work is 'industrialized'. Interestingly, Peters sees both advantages and disadvantages in the division of labour. On the positive side, it leads to specialization and this can result in high quality tuition with contributions from many different experts. On the other hand, it can also produce an alienated 'workforce', with professional teachers feeling constrained by strictly delineated roles and responsibilities. These are issues to be addressed further in Part B of this book.

The third characteristic on Keegan's list is noteworthy in giving equal emphasis to courseware and student support in promoting remote learning. One of the most crucial aspects of a distance education system is the way in which actual learning *activity* is linked to the provision of learning *materials*. Enormous amounts of time, effort and resource can be invested in the development of courseware for distance teaching. Once this package is ready and despatched to prospective students, the job of the production team, and indeed the major part of the distance *teaching* task, has been completed. Yet the process of *education* has hardly begun. Just because the students have received the learning materials, it does not necessarily follow that learning will take place. Unless the students become actively involved with the courseware, and maintain that involvement, they will simply become part of the 'non-starter' or 'dropout' statistics. The course package is rarely enough on its own. It provides the core academic subject matter; it may also provide a small measure of general advice and support of the type that will be needed by the majority of the students (e.g. help with pacing, study skills, interaction with the institution). What it cannot possibly do within reasonable bounds of complexity and cost is to give advice that might be relevant only to a small minority of those studying the course.

Additionally, there are some needs that, while they are common to all learners, cannot be satisfied other than on an individual basis – feedback on performance being the most obvious example. A particular difficulty for remote students is the fact that most learners need feedback not just from their tutor, but also from a peer group, so that there is a realistic bench mark of average achievement against which they can measure their success. It is with these kinds of interactions in mind that Sewart has drawn attention to the vital importance of what he calls 'intermediaries' in effective distance teaching – those whose main role within the system is to try to interpret and adapt that system in order to meet the needs of individuals. He expresses it thus (emphases are ours):

The *teaching* package is the raw academic pabulum of the institution for teaching at a distance; for the student *learning* at a distance there is need of another element, an advisory/supportive role. In conventional education the teacher can and often does perform the advisory/supportive role while acting as the source of academic knowledge. In a system of teaching at a distance we normally find a separation of the source of academic knowledge and the advisory/support role, the former being contained in the teaching package. The separation of these activities does not, in itself, create problems. Problems will however arise if the two elements are not balanced; if the teaching package predominates to the virtual or complete exclusion of the intermediary role; if the existence of these *two* activities is not constantly in the minds of those involved in teaching at a distance.

(Sewart *et al.*, 1983, pp. 51–2)

To be successful in fostering learning, a distance education system thus has to blend solitary learning activities with contacts that are capable of generating responses specific to each student's individual needs. Daniel and Marquis, in the title of a 1977 paper, described this synthesis as 'interaction and independence: getting the mixture right'. We shall return to this theme shortly, in the related context of open learning.

It is as a result of discussions such as this that the term 'distance education' has come to be regarded as the most suitable one in English for describing the process that encompasses both the distance teaching (course development and production) and distance learning (student support) subsystems. International practice is generally following the same trend, although French terminology has paradoxically moved away from the older *'télé-enseignement'* (based on the Greek, *tele* = from afar, rather than on 'television') to institutionalized usage of the phrase 'teaching at a distance', as in the Centre National d'Enseignement à Distance. However francophone Canada now favours the term *'education à distance'*, and the commonly used Spanish *'educatión a distancia'* and Portuguese *'teleducação'* are direct equivalents. The most common German term, particularly at the level of higher education, is *'Fernstudium'* (distance study), which seems to acknowledge the learning as well as the teaching side of the process, although *'Fernunterricht'* (distance teaching) is still in use for some technical and vocational training.

In the literature written in English, further problems arise because a number of terms have come to mean different things in different educational systems. Furthermore, some terms have been used almost interchangeably in the past but are now no longer regarded as synonymous. A detailed analysis of phrases such as 'home study', 'external study' and 'extension programmes' has been given by Keegan (1990); any reference to these terms in this book will only be in the context of particular systems within which their meaning is obvious. However, to avoid confusion it is worth clarifying our usage of two rather more general terms:

1. 'Correspondence teaching', as described in the introduction to this chapter,

was the precursor of distance teaching as it is understood today. This would still be the appropriate term for a teaching system based entirely on print or written communication sent through the postal service, but such systems are nowadays relatively rare. In our terminology, the phrase 'correspondence tuition' refers to that element of the distance education system in which communication between an educator and an individual learner occurs by exchange of letters or other written means (especially by written comments on work submitted by the student – a method of teaching that we will examine in detail in Chapter 13).

2. 'Independent learning' is a phrase that crops up frequently and in many different contexts. For example, 'independent study' is a term often used in the USA for an individualized study programme negotiated between a learner and an institution; the various courses within such a programme need not necessarily be studied at a distance, though some or all may be. In Britain the term is more often taken to imply study independent of any institution, and it is in this sense that we shall use it. There was much discussion a few years ago at the British Open University (UKOU) about the extent to which 'fostering the independent learner' should be seen as a desirable goal, by which was meant, as Wedemeyer had written a decade earlier (1977, p. 2114), 'developing in all learners the capacity to carry on self-directed learning, the ultimate maturity required of the educated person'.

3. PIONEERING DISTANCE EDUCATION: THE UK EXPERIENCE

Remote learning already has a fairly long history, and is spreading on a global scale. The range of distance education systems, projects and experiments is already vast, and constantly growing. Interested readers will find a fuller description of the role of distance education in national and international development and a more extensive catalogue of projects in an article by Rumble (1989b).

In this section we will give a brief description of the foundation and growth of the UKOU, partly because the practices of this institution are central to many parts of this book, but also mainly because of the tremendous impact it has made on the world of distance learning. Of the origins of the UKOU, Walter Perry, its first Vice-Chancellor, wrote:

> The concept . . . evolved from the convergence of three major postwar educational trends. The first of these concerns developments in the provision for adult education, the second the growth of educational broadcasting and the third the political objective of promoting the spread of egalitarianism in education.
>
> (Perry, 1976, p. 1)

In Britain in the 1960s, most 'adult education' courses were non-vocational and non-credit-bearing. Mainstream higher education catered almost exclusively for school leavers. The time seemed to be ripe for setting the media of mass communication to work in the service of education, so as to counteract elitism in educational opportunities. The UKOU was thus envisaged as an institution that would serve the needs of adult students wishing to study at post-secondary level and would admit such students regardless of their previous formal qualifications (or lack of them). The university was granted its Royal Charter in 1969, giving it the right to confer its own degrees, and enrolled its first students in 1971. Then, as now, the principal teaching medium was the specially written self-instructional text, supplemented by broadcast TV and radio and home experiment kits. In more recent years, increasing use has been made of audio and videocassettes, and various forms of computer-assisted learning. Tutorial assistance in face-to-face meetings is offered, though attendance is not mandatory; some courses have compulsory one-week, residential Summer Schools. All students are assigned to an academic counsellor and considerable stress is laid upon the importance of this aspect of the service.

The UKOU produced its hundred thousandth graduate in 1990 and, looking back now after 25 years of successful operation, it is difficult to remember the scepticism and downright hostility with which the UKOU was greeted at the outset. Perry recalled that adult education and extra-mural departments feared a take-over, academics from the mainstream universities did not believe that the techniques of distance education were appropriate for the teaching of undergraduates, or that adults without the 'standard' qualifications could study at that level, and many people considered that any initial demand for the new provision would not be sustained. Interestingly, Perry also said that it was very much a case of the prophet being without honour in his own country and described the huge overseas interest in the UKOU experiment as 'marked by an enthusiasm and an absence of the criticism so common in this country, . . . [which] did much to offset the depressing effect of the continuous flow of adverse comments at home' (p. 33). Gradually, however, the sceptics were won over, a conversion in which the imaginative nature and academic excellence of the teaching materials, the quality of the graduates and the perceived cost-effectiveness of the distance education mode all played a part.

Trow remarked in 1972 that the legitimizing of the institution was greatly assisted by the fact that it was set within a system of higher education in which common standards are rigorously applied nationwide to all degrees. While the UKOU's practices have been copied or adapted in other institutions in many other parts of the world, and initiated much research into educational strategy and techniques, there is no doubt that one of the university's most important contributions has been the credibility it has given to distance teaching. An eloquent testimony to this impact has been given by Smith (who, it should be noted, writes not from a British but from an Australian viewpoint):

The advent of the British Open University added academic respectability to distance education, lending moral support and confidence to teaching staff and administrators involved in distance education, to the students using the mode, and to the wider community that needs to have confidence in this mode of education.

(Smith, 1987, p. 30)

The UKOU's Vice-Chancellor voiced many people's hopes for the future when he wrote:

expertise in distance education is now available all over the world. The very diversity of applications . . . has created a degree of equality between institutions and countries since each must defer to the other's competence in some areas. The time is ripe . . . to work together and bring to the education of the global village the idealism that has inspired the expansion of distance education at the national level.

(Daniel, 1988, p. 58)

4. OPEN LEARNING

During the 1980s, the concepts of 'distance' and 'open' education became very much entwined. This lack of clear differentiation between what are in fact quite different ideas was probably largely due to the influence of the UKOU and other Open Universities in developing distance teaching and open learning simultaneously. Recently however, attempts have been made to define the terms more fully and to use them more precisely. Rumble's analysis quoted at the head of this chapter provides perhaps the most succinct summary:

Distance education is a *method* of education. It differs from contiguous education. The concept of open learning describes the *nature* of the education offered either contiguously or at a distance.

(Rumble, 1989a, p. 28, our emphases)

What then is the 'nature' of an open learning experience? Although the definitions of open education may have been extended or refined over the years, the three key elements in the definitions have always been

1. the opening up of opportunities as a result of the removal of constraints to learning, be these constraints practical or educational;
2. a learner-centred approach, in which individuals are empowered to take responsibility for their own learning and are encouraged towards autonomy;
3. educational flexibility, involving the use of a great variety of teaching/learning methods. As early as 1977 Coffey was defining open learning systems as involving 'the widest range of teaching strategies, in particular those using independent and individualised learning'. Today, increasing amounts of attention are being paid to these pedagogical aspects of openness.

In its initial stages the UKOU (and the similar institutions that followed it in other parts of the world) laid great stress on openness of *access*. In large measure, increased access to educational provision was brought about by the removal of practical restrictions, such as barriers of time, place and cost. In the UKOU system, one major educational barrier was also immediately set aside, namely that of entry qualifications, but other educational constraints remained. Although a range of teaching methods was used, the structure of the courses was determined by the institution, not by the students: all the students on a particular course worked through the material in virtually the same sequence, and, being tied to broadcasts, continuous assessment and examination dates, at approximately the same pace. The UKOU is just one example of a teaching system that is extremely open in some respects, but almost completely closed in others. Indeed no system is ever fully open, and very few, even the most traditional, are entirely closed. Lewis and Spencer (1986) first suggested the device of the 'openness-closure continuum' on which an educational system can be broken down into its constituents and each part analysed separately. Figure 1.1 shows just two of the characteristics that can be plotted on such a continuum, and illustrates that, as with any continuous spectrum, points may fall anywhere across the range. Various extensions and modifications to Lewis's scheme have been proposed. For example a two-dimensional graph suggested by Kember and Murphy (1990) allows administrative or logistical openness to be plotted on one axis and the degree of learner-centredness on the other.

It is interesting to keep the continuum in mind when discussing the three key elements of open education listed above. The first of these was the removal of constraints to learning. As we have seen, the early emphasis in this context

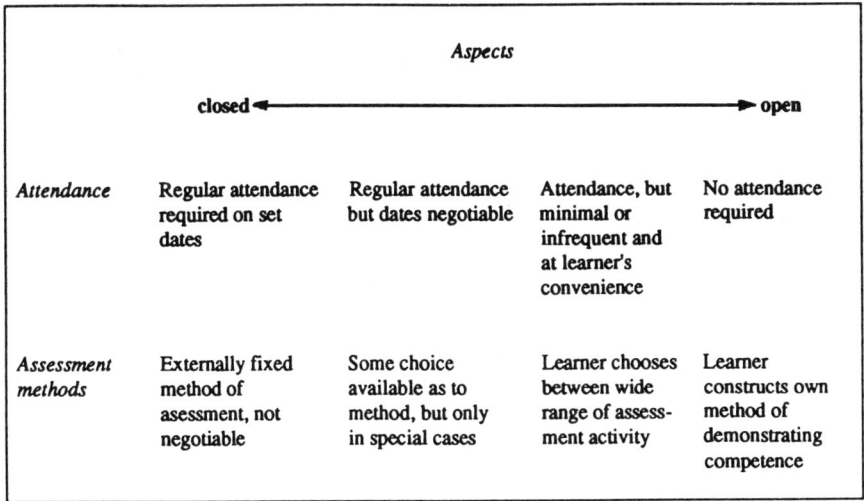

	Aspects			
	closed ◀————————————————————▶ open			
Attendance	Regular attendance required on set dates	Regular attendance but dates negotiable	Attendance, but minimal or infrequent and at learner's convenience	No attendance required
Assessment methods	Externally fixed method of asessment, not negotiable	Some choice available as to method, but only in special cases	Learner chooses between wide range of assess-ment activity	Learner constructs own method of demonstrating competence

Figure 1.1. The openness–closure learning continuum (adapted from Lewis, 1986).

(and, in Britain anyway, the funding) was on removing barriers to access, especially for adult learners. 'Open access' was seen to mean that learners should be able to avail themselves of educational opportunities, regardless of their age, social or financial status, or job (or lack of it). In practice, this was often not feasible within existing mainstream systems. The establishment of the UKOU was significant in bringing recognition that the distance education method could remove not only barriers of age and status, but other types of logistical constraints too – practical restrictions that could otherwise prove insurmountable to adult learners. Clearly access is greatly enhanced if students are not required to attend classes at particular times or places. Arguably the most open aspect of distance education is the way in which it can reach audiences excluded from mainstream provision by force of circumstance: for example the housebound, shift workers, seamen, long stay patients, prisoners and those isolated by geography. The next group of barriers to learning might be classified as educational, as opposed to logistical, but relate more to institutions than to the learning process itself. Foremost among these types of constraints are entry qualifications. Entrance requirements have to be decided for each individual course, programme or institution, and do not depend directly on the type of instruction offered by the institution. Both contiguous and distance education programmes therefore vary widely in their stance on the related issues of entrance requirements and recognition given to prior experiential (as opposed to formally academic) learning. One of the most controversial aspects of the UKOU at its inception was its admittance as undergraduates of students with no formal educational qualifications, but this lack of entrance requirements has always been a cornerstone of its policy. Looking to Britain's European neighbours, the Dutch Open Universiteit follows the UKOU example in this respect, whereas the Fern Universität imposes the same entrance requirements upon its students as the mainstream German universities. These institutions also provide examples of different degrees of openness in relation to other facets of what might be termed 'educational administration'. In matters of pacing, for examples, the Dutch OU system is far more open than its British counterpart. Dutch students may enrol at any time and then start a course almost immediately; it is also largely up to them to decide when (though not how) they will be examined. UKOU students, on the other hand, are constrained by fixed annual starting and examination dates. Within the time limits imposed by this structure, pacing of courses at the UKOU is largely determined by broadcast and assignment schedules. Radio and television transmissions are made at set times, with programmes broadcast once or at most twice in the annual cycle. Assignments must be completed by particular dates, with extensions being granted only in special circumstances. Thus in matters of timetabling, the UKOU system is almost totally closed. Nevertheless, it is appropriate to reflect at this point that the negative connotations associated with the phrase 'closed system' are not necessarily justified. For example, there are many who would argue that the tight pacing of UKOU courses has definite and demonstrable advantages over more 'relaxed' schedules in terms of the throughput and success rate of students.

The second key feature of open education is that it should be learner-centred. Again, this condition implies the removal of constraints – here constraints on course content and structure. In a totally open system, students would have freedom of choice not only over pacing, but also over learning objectives, sequencing, teaching methods and assessment. Students would thus define their own objectives and negotiate a syllabus with the institution or teacher. An example of a higher education programme organized along these lines is that offered by Empire State College (one of the colleges of the State University of New York). As described by Granger (1990), the Empire State system involves each student in a process of individualized educational planning that takes account of prior formal (academic) study and/or experiential learning, identifies academic and professional goals and then formulates a programme that is the basis of a series of 'learning contracts' with the College. The emphasis in this scheme is on openness of curriculum, rather than on openness of access, although the latter may be available as well. Pacing is decided mainly by the student, with the learning contract drawn up accordingly. Because in the Empire State model the student's ultimate reward is a degree, the content and attainment levels within each programme are to a large extent set by the wider academic community. Nevertheless, the fact that each programme is tailored to an individual means that the student will be clear about the programme's structure and coherence, and satisfied that its content is relevant to his or her needs. In a quest for more open forms of education, the learning contract concept could be adapted to learning over a wide range of academic levels (although it is probably more suited to adult learners than to children), with the institution's side of the contract being fulfilled by distance, contiguous or mixed-mode teaching, as appropriate to the particular circumstances of the learner. However, it is when the actual teaching methods are finally addressed that the ideals of openness become increasingly difficult to apply.

This is the point at which the three key features of open learning listed above finally all come together. The nature of the teaching has to be such that barriers are not created by lack of flexibility in the mode of delivery and assessment. If openness is to be maintained, there must be a choice of routes and learning strategies by which the educational objectives may be reached. Individual learning styles have to be catered for, and their individual potential realized. A multiple media approach can thus contribute to openness, especially if a certain amount of redundancy is built in. Complete redundancy, such that the learner can choose to concentrate on one medium and neglect another without jeopardizing his or her attainment of the overall objectives, is rare (largely because it is expensive!). However, partial redundancy, with similar material covered in different ways and perhaps at different depths by different media is not unusual. Here, as elsewhere, openness is not an absolute but a relative term. Courses that include alternative or branched pathways are more open than those that do not. Project-based courses are more open than syllabus-based courses. Continual testing throughout the duration of a course is a more open form of assessment than a single examination at the end of the course. There are often, however, particular difficulties for courses leading to formal qualifications in trying to reconcile rather open types of assessment with certification. Part B of this book examines the various problems of course design for both distance and open learning courses in more detail.

A course that is open in the educational sense is one that employs a range of teaching strategies. The converse is not necessarily true: not all multiple media courses are educationally open. It is the nature of the teaching strategies and not their number that fixes the place of a course on the open–closed continuum. The distance teaching institutions use, in different combinations, virtually all the available communication media, but this fact does not of itself lead to educational openness. Some of these media (set texts, TV and radio broadcasts, audio- and video-tapes) are highly programmed and their dominance can result in tightly structured, inflexible – in other words, closed – courses. In the worst cases, as Rumble (1989a, pp. 31–2) points out, this 'control of the curriculum and the isolation of the student body . . . make distance education a ready vehicle for totalitarian control of education'. In more democratic regimes, it has also long been recognized that the capital investment a distance education institution makes in its courseware can be a drawback as well as a strength. One danger is that too much of the energy of the institution may go into the creation of a course package, and too little attention be given to the 'service' aspects of the enterprise. A balance may be restored if the package also involves media that permit real two-way communication (correspondence, telephone, electronic mail or computer conferencing), but the degree of openness conferred by such media depends on the extent to which they are used for genuine dialogue and negotiation. An exchange of letters or a phone call through which a student seeks clarification of a set text and the tutor supplies an authoritative response does not constitute a particularly open interaction. New developments in computer conferencing, on the other hand, may be valuable in opening up educational opportunities in a remote learning context since, like seminars in a contiguous learning context, they permit student-to-student contact and peer teaching. The messages of the structured parts of the course package may thereby be individualized, and students encouraged to assume more responsibility for their own and one another's learning. It is in the development of learner autonomy and self-reliance that the benefits of open learning are increasingly seen to lie. Surely this is right. Openness of access to educational provision is important. So is receptivity to new teaching techniques. But in all these discussions of institutions and systems, one must not lose sight of the ultimate measure of success of any educational endeavour – the nature and quality of the learning experience that results. As Temple (1988, p. 116) reminds all educators, the real power of open learning is not just in removing constraints of time, place and curriculum, but in 'removing the barriers which preconceptions about ability and attainment and the ways in which these are manifested raise against the fulfilment of individual potential'.

5. THE WAY FORWARD

Over a decade has passed since Pagney (1980, translated 1983) pointed out that the whole educational process would be enriched if distance teaching were fully recognized as being complementary to, rather than in competition with, mainstream teaching. He looked forward to such recognition as the coming-of-age of distance teaching, when it would 'shed its marginal image once and for all, and be accepted as an integral part of the educational system' (1983, p. 157). Today, there are encouraging signs that distance teaching is indeed approaching this coming-of-age, with many of the philosophies and practices of distance and open education being increasingly applied in the mainstream institutions.

Smith and Kelly (1987), in an analysis based chiefly on the Australian experience but equally applicable elsewhere, have identified three principal ways in which distance education and mainstream education are converging:

1. an interchange of teaching methods, mostly, but not exclusively, in the direction of techniques derived from the practices of distance education being incorporated into mainstream systems;
2. an intermixing of the clienteles for the distance and mainstream sectors;
3. a movement, in both distance and mainstream education, towards more open forms of learning.

The growing awareness within the mainstream sector of the methodology of distance education, and the increasing pool of people who have attained qualifications by distance/open learning, tend not only to accelerate the process of convergence, but also to push it into new areas. The potential of distance education techniques is now being recognized in many fields – for the initial and in-service training of teachers, for industrial training, for professional updating. Resource-based teaching and open learning are spreading into both secondary and primary schools. It is perhaps in the continual widening of the scope of distance education that the most interesting future prospects lie. It is debatable whether the tremendous growth of distance education in the 1970s and 1980s would have occurred without the proliferation of new communications and information technology that characterized the period. Distance education has its origins in the provision of printed material and correspondence lessons sent through the postal system, but in the developed world it has moved far beyond this, firstly into the use of broadcast media, audio materials and the telephone network, and more recently into the many forms of computer mediated communication. Even in less developed countries, radio and (to a lesser extent) television have played important educative roles, helping to overcome problems of low literacy levels and poor postal services. Much of the innovative thrust of distance education has come from a willingness to explore all available methodologies and to make use of technological advances as they occur. In the past two decades, the flexible attitudes of most distance educators to styles of tuition have resulted in the development of many new teaching techniques. Some of these approaches have been driven directly by the wish to apply new hardware or software within teaching/learning packages. Others have evolved, at least in part, because of computer mediated research into a variety of educational issues. The computerized student record and assessment systems of the large distance teaching universities have played a major role in the evaluation and on-going development of those institutions' teaching programmes. Computer-marked assignments and questionnaires allow objective feedback to be gathered with comparative ease from very large numbers of students. Such data have been widely used, both for 'market-survey' investigations into student needs and perceptions, and for research into the educational effectiveness of different teaching strategies. Institutional policy and course design have been greatly influenced by such studies.

In these respects, educational thinking in the 1980s was far more radical and innovative in the distance teaching institutions than in the conventional ones, particularly at the tertiary level. It is with this background that the convergence of distance and mainstream practices is now beginning to take place. Distance education is thus poised to act as a major agent for change in other areas of education and training, enhancing their awareness of the possibilities of multiple media approaches and of new technologies. For distance education to continue in such a role, however, it must retain its flexible attitudes, further developing and implementing new methods.

The final part of this book (Chapter 15) highlights the areas that seem the most likely to offer fresh perspectives in the next decade. On the operations front, the production of course material is being revolutionized by a widespread conversion from conventional to electronic publishing. On the technological front, new and forthcoming hardware and software promise the rapid development of various types of conferencing network, and herald increasingly sophisticated graphics and simulation packages. All these advances will have educational applications. The movement towards open forms of learning should be particularly well served by the increased availability of computer networking, and by the current interest in less authoritarian forms of computer-aided instruction.

2

OPENING UP SCIENCE

[Science should be more than] a reserved compartment in which only specialists have a right to travel.
(HMSO, *Science in Secondary Schools*, 1960, p. 5)

1. INTRODUCTION

The slogan 'science for all', which first emerged from a UNESCO working party in 1983, has been seen in many countries as the embodiment of a new approach to science education aimed at improving the scientific literacy of the public at large. The National Science Foundation (1983) and the American Association for the Advancement of Science (1989) both published plans in the mid-eighties aimed at improving science education in the USA; the Science Council of Canada and the Curriculum Development Centre (1988) in Australia produced statements with similar titles and aims. In the UK, The Royal Society boldly declared in 1985 'Science *is* for everybody'; an organization devoted to studying and fostering the public understanding of science (COPUS) and a journal entitled *Public Understanding of Science* were established in the early 1990s.

> The importance of a scientifically and technologically literate population is being emphasised in all countries, since it is recognised that specialist scientists and technologists cannot operate without a knowledgeable supporting society.
>
> (Harlen, 1993, p. 126)

Many authors have commented on particular aspects of the 'science for all' approach; Fensham's reviews (1988, 1992) provide an especially interesting analysis of the curriculum development that has resulted. Although the 'science for all' movement has been mainly addressed to the compulsory education sector (primary and secondary), many of the points on which it is based apply equally to the programmes of tertiary, adult and open education institutions. In particular, whatever the level, the choice of curriculum content is crucial. Fensham, for example (1988, 1992), has argued that content in science courses should fulfill two essential criteria: it should have societal meaning and usefulness to the majority of learners, and it should allow learners to share in the wonder and excitement that has made the development of science such a great human and cultural achievement. Other issues that must be addressed relate to the provision of gender-inclusive learning materials (Fensham, 1993; Harlen, 1993), differentiated access for students with special educational needs (Purnell, 1993) and a generally more diversified range of opportunities for potential science learners. The open and distance education sector is well placed to satisfy all these criteria, and hence to recruit new adult audiences for science courses.

2. TEACHING AND LEARNING SCIENCE

The increasing number and sales of 'popular science' books in the last decade may be taken as evidence of a growing public interest in or at least awareness of science (see, for example, the essay by Rodgers, 1992). Such books, however, tend to concentrate on a rather narrow range of (often curiously 'advanced') topics, and to be largely descriptive in nature. They may spark interest in potential science learners, but are in no way a substitute for formal instruction. The following quotation from *Primary Science: Why and How* gives a good working definition of what we mean by 'science' in the phrase 'teaching or learning science'.

> One definition of science is that science is what scientists do. Scientists actively confront nature with questions, seeking patterns in what they observe so they can make predictions and possibly use and control aspects of it. The ways in which scientists proceed with their observations, questions and experiments involve processes aimed at achieving reliable, reproducible and objective answers, with the result that scientific knowledge is verifiable. So science is not only the body of knowledge that scientists have accumulated but also the ways in which they acquire this knowledge. These are the 'content' and 'process' dimensions of science so much debated in the context of appropriate science teaching.
>
> (Open University, 1985, p. 8)

The relationship between the content and process aspects of science education will be further explored in Chapters 3 and 5, as we discuss theories of learning and skills development. But the distinction is a crucial one for curriculum and instructional designers in deciding whether they want to produce courses *about* science or courses *in* science. Most science educators consider that the process dimension excites the curiosity and attention of learners:

> There is a strong case for arguing that process-based science is likely to be more stimulating . . . than the presently predominating content-based approach. Further, because it conveys an image of science as imaginative and created by the human mind, it is more likely to be seen as interesting and relevant.
>
> (Harlen, 1993, p. 127)

On the other hand, there is an obvious tension between the requirement of any technological society to train future scientists and technologists in the details of their subjects and the aim of producing an informed and scientifically literate public. Fensham and Harlen, among others, have pointed to the dangers of separating these two categories of learners. The fact that intending scientists have to learn 'real science' does not lessen their need for an appreciation of either the societal relevance of science or its moral and ethical dimension. Conversely, it is impossible to draw the public into the decision-making process about issues such as pollution, genetic engineering or nuclear power if all knowledge about these areas rests only with a small constituency of élite professional scientists.

Different countries have resolved this tension in different ways. Some have opted for a broad curriculum of integrated science in compulsory schooling up to a certain level. Others have streamed or modular programmes that separate the intending science specialists, usually at some point in their secondary education. Before the advent of the 'open', distance-teaching universities there was very little opportunity for direct entry to science or technology courses at the tertiary level for students who had not completed the necessary selection and preparation in their secondary schooling. And as Fensham has pointed out,

> The two tasks of selection and preparation are usually associated. That is, the content of science education deemed a suitable preparation for the science disciplines in tertiary education has also turned out to be a useful selective device since comparatively few students learn it successfully. Whether this low success is due to inherent difficulty in the content of science, or whether it is due to lack of interest among students is another issue.
>
> (Fensham, 1993, p. 112)

It is this issue that we will explore in the next section.

3. WHY SCIENCE IS DIFFICULT TO LEARN

There is a public perception that science, like maths, is hard to learn. Most schoolchildren in the Western world receive at least some exposure to science at school, yet many adults are unable to answer correctly questions relating to basic scientific facts (see e.g. Lucas, 1987). Public understanding of the history of scientific achievement and processes of science seems equally deficient. In a survey of 2,000 adults undertaken in 1985 by the British magazine *New Scientist*, only 45 per cent of the sample thought that science did more good than harm, and a further 38 per cent considered that the good and the harm cancelled out. Astonishingly, 27 per cent of men and 45 per cent of women questioned failed to name correctly even one important scientific achievement which occurred after the Second World War. Although the 'science for all' movement is encouraging more people to learn science, it is doing so in a culture in which science is not consistently highly valued. So why is it that much formal teaching of science apparently fails to put across the basic concepts, methodology and value of science?

Millar (1991) has suggested four main reasons for science being hard to learn:

1. learners do not feel that the efforts they must make to learn science are matched by a sufficently large payoff;
2. an understanding of science involves continual reconstruction of meaning, and cannot be achieved by rote learning of a fixed body of knowledge;
3. many learners are confused or alienated by certain aspects of the nature of science;
4. science is abstract.

Item (1) on this list relates to the perceived relevance of an understanding of science for people's everyday lives, to which we have already alluded in the two previous sections. Science educators have not on the whole made a good job of identifying or demonstrating what part of science is of value for *all* learners, nor why it is of value. In many cases this has led to an overburdening of syllabuses, which is counter-productive in terms of learning outcomes.

We will discuss item (2) – the requirement for reconstruction of knowledge – more fully in Chapter 3, but the underlying point is that, in order to switch from everyday perspectives of phenomena to a scientific understanding of those phenomena, the science learner has to restructure prior conceptions; such changes of gestalt constitute a more demanding form of learning than mere accretion of knowledge.

Millar has suggested that item (3) arises from a discrepancy between the attitude of many science educators, who often present science essentially as an algorithm for obtaining knowledge of how the natural and physical world works, and the true nature of science as a body of understanding consensually accepted within the practising community. He believes that

> Science education needs first to acknowledge and then to address the tension between the 'openness' of science as first-hand investigation and the 'dogmatism' of science as a consensually accepted body of knowledge.
> (Millar, 1991, p. 70)

If this tension is not resolved at the instructional level, there is, as Russell (1983) has pointed out, a danger that learners can be required to accept conclusions without appreciating the underlying evidence. This can give them a very distorted picture of the nature of science, presenting it as a collection of often unrelated or useless facts, and hiding its overarching purpose. Under such circumstances, it is hardly surprising that many learners are confused, demotivated or alienated.

Item (4) encapsulates an issue to which many commentators refer. For example, Millar points to

> compelling evidence that the basic units of understanding are concrete examplars, not rules or syllogisms . . . This suggests that the knowledge is stored not as a syllogism (a formal logical rule) but as concrete examples of the successful application of that rule. When we use the abstract rule, we check it against our recall of particular instances and not the other way round.
>
> (Millar, 1991, p. 71)

Shayer and Adey (1981) have gone so far as to suggest that attempts to teach real physics to children are doomed to inevitable failure, because an understanding of physical principles can only be based on formal operational thinking, which in turn demands that the learner should have reached a certain stage of cognitive development. It has been shown that in fact many students do not reach this stage even by the end of their secondary education, so it is not surprising that they cannot cope with the abstractions in which much science is expressed. Related to this is the requirement to deal with a variety of representations of real situations. Johnstone (1991) has highlighted this problem, drawing particular attention to the difficulties many students experience with multi-level thought; some examples are illustrated in Figure 2.1.

Another very obvious source of difficulty for science learners is the language in which science tends to be couched. For one thing, it is full of technical terms and jargon words. Interestingly however, it has been shown (see e.g. Cassells and Johnstone, 1983) that new technical terms cause fewer problems for learners than words that are in everyday use in one sense but have a different and precise meaning in the context of science. Johnstone gives the following example:

> 'Volatile' has left the realm of science with its meaning of 'easily vaporized' and gone off into common speech where it is applied to markets, people, countries and hostile situations. It then filters back into science with meanings that [to students] do not seem out of place . . . but which make a nonsense of a science discussion. A 'volatile compound' is understood as a 'flammable, explosive, unstable and dangerous compound'.
>
> (Johnstone, 1991, p. 80)

We will return to this issue in the next chapter, when we take up in more detail the point about science learning as a reconstruction activity. Difficulties with language also tie back to the abstract nature of science. The language of science, unlike that of, say, history, is removed from the language of everyday intercourse, and this further contributes to its perceived abstraction. This problem is compounded by the fact that in many scientific disciplines the abstract language is coded in symbolic form – the language of mathematics.

The claim that science is hard to learn is really a statement that science educators are not getting the message across. Part of the problem, to which we

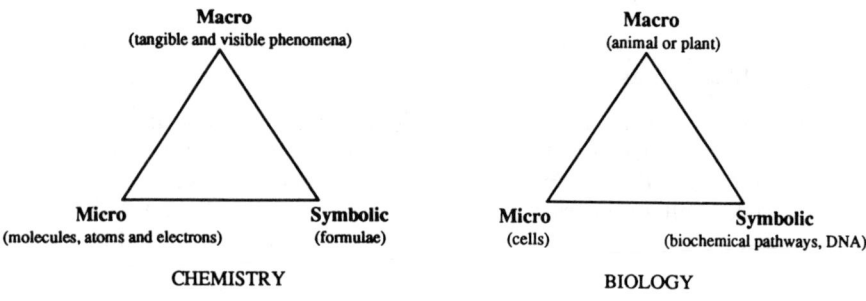

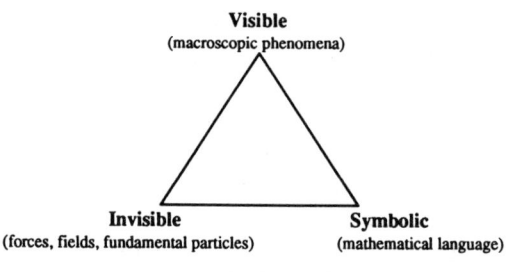

Figure 2.1. Diagrams illustrating 'triangles' of thought levels for several scientific disciplines (adapted from Johnstone, 1991, p. 78). Some good science can be done at one level only (usually the macro), or at two (e.g. classical thermodynamics which involves only the macro and symbolic levels). However, much science teaching is done 'within the triangles', where all three levels interact. Students' difficulties in juggling the three levels simultaneously may account for many of the problems they have in learning science.

---

have alluded several times already, may lie with the very nature of the message; instructors need to pay more attention to demonstrating the significance and relevance of that message. As mentioned briefly in this section, some of the faults may also lie with instructors' failure to appreciate the ways in which science learning is achieved; this is a subject we will explore in more detail in the next chapter. But there may also be major problems with the transmission systems used to convey the message, and these are of particular concern in open and distance education.

4. THE CHALLENGE OF OPEN OR DISTANCE EDUCATION IN SCIENCE

In the early days of the UKOU, Kaye and Pentz – respectively an educational technologist and the original Dean of the Science Faculty – gave an analysis

(Kaye 1973; Kaye and Pentz, 1974) of the specific problems associated with the design of the institution's first science courses. They referred to some of the difficulties already discussed above, which apply across a broad spectrum of science teaching and learning, but may be exacerbated in the case of remote students. They also pointed to problems inherent in a system that offered science instruction at undergraduate level to those without prior qualifications. Language issues, some aspects of which have already been mentioned in the previous section, are one example. Any course or institution with open entry policies will have a very heterogeneous body of students, particularly at lower levels. These students need to become comfortable in talking about science, and even comparatively trivial difficulties, such as uncertainty about the pronunciation of a technical term, can create barriers to progress. Students' misunderstandings about the meaning of particular words in a scientific context are also more difficult for instructors to pick up in a remote teaching environment than in conventional classroom or lecture situations.

Another area of language difficulty can arise out of modular systems, in which students can range quite widely across different areas of science. Scientists themselves often have problems communicating with colleagues from other scientific disciplines, as much because of differences in meaning attached to certain terms as because of unfamiliar jargon. The construction of glossaries associated with individual courses can offer partial solutions for learners in such situations, but even this can be difficult in the less 'settled' subject areas, where the meaning of terms is still shifting.

Modular course structures, especially those with open entry systems, present instructors with particular problems in respect of subjects like science that have a strongly sequential or hierarchical knowledge base. An understanding of many concepts in science requires a grasp of a significant number of lower level concepts. Thus, before physics students can analyse a collision, they need to have understood the concepts of mass, velocity and speed, momentum, energy and conserved quantities, and each of these concepts can be unpacked in their turn to produce an even lower level list. The syllabus for any science course is in fact an elaborate edifice based on assumed prior knowledge.

In very 'open' modular systems the problem of hierarchies and prerequisites needs to be addressed not only at the level of individual courses, but also across the entire spectrum of courses. The UKOU, for example, has always attempted to allow for the diversity in its students' entry behaviours and aspirations by stressing its open entry policies; these impose no entrance qualifications and permit students an extremely wide freedom of choice in the construction of their degree profiles. Nevertheless, within this free choice system, considerable care is taken to warn students of the pitfalls of eccentric degree profiles and to provide them with advice about sensible combinations and ordering of courses. Although many UKOU students do elect to concentrate mostly on one discipline, a study (Ross and Scanlon, 1989) of UKOU 'science' graduates has shown that the freedom to mix and match courses is widely valued. This diversity of degree profiles is hardly surprising in view of the range of previous experience and aspirations that UKOU students bring to their studies in science: some are updating or upgrading earlier science qualifications, others are embarking on science for the first time, but with intentions of pursuing the field in either teaching or research, still others are doing courses simply for their own interest. One measure of the success of the UKOU's modular system is the extent to which it successfully fulfills the requirements of such a diverse student body.

Another issue highlighted by Kaye and Pentz (1974), and not unrelated to the subject of prerequisites for science study, concerns the development of skills, especially mathematical skills. The opportunities afforded by distance education may bring new audiences to science courses, but these audiences seldom come equipped with the necessary mathematical skills for undergraduate level science; with open entry, this problem is even more severe, particularly at the foundation level. In designing courses, instructors need to be extremely sensitive to the demands on students who may be grappling for the first time with mathematical formulations as well as trying to understand new scientific concepts. In Chapter 5 we will argue that skills development needs to be built into the fabric of courses from the outset, and that the weaving in of process-based material must be an integral part of course design.

Constituting a different class of skills and processes, but equally vital to the nature of science, is practical work. The difficulty of providing adequate practical experience has often been seen as one of the chief stumbling blocks for science instruction at a distance. This is such a major issue that the whole of Chapter 10 is devoted to it.

In Part B, we aim to show how science educators have risen to the challenge of producing open science courses for remote students at undergraduate level. Many of the examples we will use derive from the UKOU, which has had a quarter of a century's experience in science education at a distance. The statistics of this institution certainly indicate that open and distance education in science can be attractive and effective: by the early 1990s a number equal to around 10 per cent of full-time British 'fresher' science undergraduates annually began studying science with the UKOU, and it has been estimated (Scanlon *et al.*, 1993) that over the first twenty years of the university's existence a total of 53,000 adults had acquired scientific skills and knowledge by studying its science foundation course.

However, before the detailed discussion of courseware which makes up most of Part B, we will consider some of the general principles underlying the design of distance education courses in science. The next chapter therefore combines perspectives of instructional design with the findings of research into the ways in which science learners develop their understanding of the subject.

3

INSTRUCTIONAL DESIGN

Technology provides us with powerful tools to try out different designs, so that instead of theories of education, we may begin to develop a science of education. But it cannot be an analytic science like physics or psychology; rather it must be a design science more like aeronautics or artificial intelligence. For example, in aeronautics the goal is to elucidate how different designs contribute to lift, drag manoeuvrability, etc. Similarly, a design science of education must determine how different designs of learning environments contribute to learning, cooperation, motivation, etc.

(Collins, 1993, p. 15)

1. INTRODUCTION

This chapter will deal with various perspectives in psychology relating to learning theories and their application to the design of instructional materials. We focus particularly on aspects that relate specifically to science education and to the design of course materials in distance learning.

It is an assumption of the instructional design approach that in fact instruction *can* be designed to some good effect. This belief is often attacked by those who worry that such an approach leads to prescriptions which stifle individual creativity among teachers. On the other hand, instructional designers including e.g. Romiszowski (1981, 1984) argue strongly that a systems-based approach to instructional design need not be confined to a series of mechanistic algorithms, but can offer the framework for heuristic strategies adaptable to a wide variety of teaching and learning styles.

At one time instructional designers could be criticized for simply promoting the development of particular student behaviour, but now there is an emphasis on developing the cognitive structure that enables the student to behave appropriately. We will comment on this change of focus in the aims of instructional designers in general in the next section. Section 3 reviews recent work on the ways in which students learn science, and Section 4 looks at the implications for the design of distance teaching materials for science courses.

This chapter concentrates on the general features that influence instructional design decisions, but does not provide descriptions of how such decisions might be implemented. Detailed discussion of the selection and use of particular instructional media and techniques is the main focus of Part B.

2. NOTES ON THE HISTORY OF INSTRUCTIONAL DESIGN

Instructional design has been considerably influenced by work on the psychology of learning. The ideas of Dewey (1916) and Thorndike (1913) (both professors at Columbia University in New York at the beginning of the century) played a particular part in this development. Thorndike, a professor of psychology, investigated the connection between stimulus and response, ideas later developed by Skinner. Thorndike coined a number of 'laws of learning' as part of a theory, popular among psychologists of the time, known as connectionism. This theory of learning was based on the assumption that a connection would be established between stimulus and response when a stimulus produced a satisfying response in a particular environment. The law of exercise or repetition said that the more often a stimulus is repeated, the longer it will be retained. The law of effect said that a response is strengthened if it is followed by 'a satisfying state of affairs', and weakened by 'an annoying state of affairs'. Thorndike investigated these laws mostly with animals, but went on to study human instruction. For example, he considered what the consequences of the existence of stimulus–response bonds were for the study of arithmetic. The principle of reinforcement, indeed the whole behavioural approach to the design of instruction which lay behind the work of Skinner, Pressey and others on programmed learning, can be traced back to Thorndike. On the other hand, Dewey, a professor of philosophy, believed that the relationship between stimulus and response was more complex.

> To learn from experience is to make a backward and forward connection between what we do to things and what we enjoy or suffer from things in consequence. Under such conditions doing becomes a trying; an experiment with the world to find out what it is like; the undergoing becomes instruction-discovery of the connection of things.
>
> (Dewey, 1916, p. 140)

His view of learning involved the environment and he advocated problem solving, laboratory work, scientific method, and the use of projects as some of the most productive methods of instruction, stressing the importance of concrete experience and active responses, rather than simple practice and feedback which are ideas from the behaviourist tradition. Dewey was influential in developing the notion of discovery learning. The chief present day exponent of discovery learning, Bruner, developed a model of instruction which was based on discovery methods (Bruner, 1960). Bruner's later theory of instruction (1966) concentrated on the internal thought processes of the learner and stages of intellectual development. He describes the learner as moving through three levels of representation of his or her knowledge: by action, image or symbol. Bruner calls these levels enactive, iconic, and symbolic. The enactive level deals with the direct manipulation of materials, the iconic deals with the manipulation of mental images of objects and the symbolic with the manipulation of symbols alone. Bruner summarized (1961, p. 26) what he saw as the main hypothesis about the benefit of discovery learning as follows: Practice in discovering for oneself teaches one to acquire information in such a way that makes that information more readily viable in problem solving.'

The problems with discovery learning are that, taken to its extreme, it can appear that the learner is being abandoned to teach him or herself, and it can be inefficient in terms of time or effort for the student. An adaptation of the extreme version of discovery learning designed to deal with these problems by giving students more support is called *guided* discovery learning. This style has been given particular prominence in recent years although a review of the effectiveness of discovery learning methods recommended the guided discovery approach as long ago as 1966 (see e.g. Shulman and Keisler, 1966).

In contrast, Ausubel (1968), another contemporary instructional designer, developed a model of instruction based on expository methods and cognitive structures as well as a theory of learning from which he devised most of his instructional development ideas. He, unlike many others, does not believe that discovery learning approaches are superior to other instructional techniques:

> The proposition that all people must discover for themselves every bit of knowledge that they *really* wish to posses is, in essence, a repudiation of the very concept of culture.
>
> (Ausubel, 1968, p. 531)

His basic claims are that activity is not necessary for learning; learning can be effective even when purely verbal methods are used, and meaningful verbal learning occurs only when new ideas are incorporated into a structure of thought that has already been established by previous learning.

The view that what a person can learn depends on what he or she already knows was also shared by Robert Gagné, who combined ideas from a variety of viewpoints (behaviourist, humanist, etc.) into a theoretical approach to the design of instruction. With his collaborators at Florida State University he has written over a period of 20 years about his views on learning and instruction (1965, 1985). In his first book he identifies eight different types of learning and arranges them in a hierarchy from simplest to most complex. These range from signal learning to problem solving. Gagné associates each type of learning with a particular type of appropriate instruction. His model of learning stresses that mastery of higher order learning depends on success with lower level learning. In principle any task can be analysed in terms of all the steps in learning which must precede the final step, though this analysis is complex even for easy tasks. So Gagné's work is a useful reminder that by breaking down tasks into smaller steps success can be achieved. Also, Gagné presents two alternative ways of learning higher order intellectual skills: the expository strategy and the guided discovery approach. The expository approach involves definition and presentation of examples, as already mentioned, while in the guided discovery approach the task is a new problem, the solution of which is promoted by hints and learning questions.

Ausubel sees the learner as having a set of pre-existent ideas which provide anchors for new information. As a result, he considers that the teacher's role is to identify the structure of the subject matter to be learned, and to transmit this in such a way that the structure can be related to the student's own cognitive structure. In his 1968 book, Ausubel formalized these ideas to define two strands of teaching:

1. progressive differentiation, whereby 'the most general and inclusive ideas of the discipline are presented first and then progressively differentiated in terms of detail and specificity' (p. 189).

This connects to his tactic of the use of advance organizers (see Chapter 6) and is in direct opposition to Gagné's advocacy of a preferred instructional route.

2. integrative reconciliation, or 'recombination of existing elements of cognitive structure' (p. 124).

According to this, previously learned concepts are modified by reconciling them and integrating them with new ideas. As will be discussed in this chapter, different students start from different cognitive structures. Alternative instructional sequences for students taking the same course would seem to be one consequence of this. As Ausubel says,

> If we had to reduce all of educational psychology to just a single principle we would say this: 'Find out what the learner already knows and teach him or her accordingly.'
>
> (Ausubel, 1968, p. 373)

Another focus of interest in instructional design is the idea that goals of learning should be explicitly identified. The need for educational objectives to be expressed was the subject of discussion from early in the century, and in the 1930s Tyler developed the concept of behavioural objectives.

> Each objective must be defined in terms which clarify the kind of behaviour which a course should help to develop among students. This helps to make clear how one can tell when the objective is being attained, since those who are reaching the objective will be characterised by the behaviour specified.
>
> (Tyler, 1934, quoted in Henderson and Nathenson, 1984, p. 67)

Bloom (1956) of the University of Chicago was the chairman of a committee, set up after a meeting of the American Psychologists' association in 1948, which met annually from 1949 until 1953 to work on a taxonomy of educational

Reigeluth	Bloom	Merril	Ausubel	Gagné
Memorizing information	Knowledge	Remember verbatim	Rote learning	Verbal information
Understanding relationships	Comprehension	Remember paraphrased	Meaningful verbal learning	-
Applying skills	Application	Use a generality	-	Verbal intellectual skills
Applying generic skills	Analysis, Synthesis, Evaluation	Find a generality	-	Cognitive strategies

Figure 3.1. Reigeluth's four major types of learning, with their associated terms in the descriptions of other theorists.

objectives. The committee members were all college or university examiners who considered a taxonomy of objectives in the cognitive domain would be useful, at least in higher education. The terms they used to classify knowledge were: comprehension, application, analysis, synthesis and evaluation. Mager later emphasized the need for behaviourally stated objectives and in 1962 produced a book which actually showed how to formulate them. Although reviews of research outlining the problems with behavioural objectives have been published (Macdonald-Ross, 1973; Melton, 1978, 1982), there are also benefits from their use, which has been widespread. To some extent they are a main component of what has become known as the systems approach to instructional design.

It could well be argued that theorists in this field have a considerable talent for inventing different names for the same things. Reigeluth (1993) believes that there are in the cognitive domain four major types of learning which require very different methods of instruction. He calls these: memorizing information, understanding relationships, applying skills, and applying generic skills. Figure 3.1 shows how these terms map on to those of other theorists, whose work we have already discussed: Ausubel, Bloom and Gagné, and Merril whose work is related to Gagné's.

Domain specific content can be acquired from any of the first three types

of learning. Concepts can be memorized, or can be understood, or their relationship with the other knowledge the learner has can be applied. The fourth
kind of learning which Reigeluth calls 'applying generic skills', but is more often
described as 'developing transferable skills', is domain independent and takes
longer to acquire. We will return to the issue of skills in Chapter 5. Most work
in instructional design has been on prescriptions for memorizing information or
applying skills. What work that has been done on the acquisition of meaningful
learning has been on the development of a descriptive learning theory rather
than prescriptive instructional theory.

Some influences on instructional design have come from unlikely sources.
Piaget's work on cognitive development is very famous. Originally a biologist,
he became interested in children's development in 1920 while helping with the
administration of routine standardization of intelligence tests. This interest led
to a forty year study of individual children as they worked on selected tasks.
Most of these tasks involved the manipulation of physical materials; from them
Piaget acquired information on how children develop their understanding of the
physical world and how they structure their ideas about such concepts as time,
space, matter and motion. He himself wrote comparatively little addressed
directly to education. However he supplied a Piagetian view of the child which
has been enormously influential, not only in projects determining science
curricula at the school level, but also as a kind of touchstone against which
ways of encouraging science learning at all levels have been judged. Piaget's
view of human beings as active, adaptive, goal-seeking, problem-solving
organisms, and his view that children learn naturally through interaction
with the environment, has probably influenced education as a whole more
than any other part of his work on children's intellectual development.

Some writers (e.g. Henderson and Nathenson, 1984) have speculated that
there is a link between the work of Piaget and Bruner. While Bruner originally
applied his three stage model to delineate the course of cognitive growth in the
age range approximately corresponding to Piaget's stages, it may be that adults
have predispositions in favour of one or another of Bruner's modes of learning.
Thus, a learner with a predisposition to the iconic mode would learn better
from visual media such as diagrams or films than from, say, audiotape; this
is an argument for the desirability of providing alternative learning pathways
for any subject – in effect an argument for multiple media teaching. This idea
has not been tested however, and Piaget's legacy remains esentially separate
from Bruner's. The idea of the importance of prior conceptions in shaping
future learning has been much applied to the study of how science is learnt
and what strategies can be used to support students' learning. This will be
discussed further in Section 3.

Bruner's more recent writing (1986) has been influenced by the Soviet
psychologist Vygotsky who worked in the early part of this century but
whose work only recently became available in the West. Vygotsky's ideas
overlap with Piaget's in that he believes in the importance of experience, but

emphasize also the role of language and culture in learning. Both Vygotsky and Bruner believe that instruction plays a very significant part in children's development and see the procedures, practices and concepts that children are taught as cultural inventions created by people in certain social contexts. Bruner (1986) defines culture as the implicit semi-connected knowledge of the world, from which, through negotiation, people arrive at satisfactory ways of acting in given contexts, and considers social context to be a key part of the process by which people learn. Vygotsky (1978) had argued that an individual's higher cognitive processes are formed in structures that are transmitted to the individual by others through speech–social interaction and co-operative activity. For example 'others', perhaps teachers or the institution in which learning takes place, provide the learner with the conditions to help observe regularities and structure in his or her own experiences and thus aid the acquisition of knowledge. Teaching can 'scaffold' the learner's activity, allowing it to go beyond the level of performance that could be achieved unassisted.

Those who concentrate on the knowledge that learners bring can be classed as 'constructivists', but they often focus only on the cognitive aspects of such learning. The contribution of Bruner and others is to emphasize the importance of social interaction. A *social* constructivist perspective of learning combines the idea that students are agents in their own learning, with the idea that the learning context is also important in terms of the interactions with others involved in the situation and the tasks which students are set. In any situation students draw on a variety of experiences to help them make sense of it. Past experiences, together with the students' views of their relevance and usefulness, will all influence the way in which tasks are interpreted. An important idea here is the role of active learning in developing useful robust knowledge (Brown, Collins and Duguid, 1989). This is particularly important for science teaching, as learners of science need to be involved in developing both their procedural and conceptual understanding by appropriate actions.

The particular aspect of social constructivism referred to in the paper by Brown *et al.* above is known as 'situated cognition'. This grew out of dissatisfaction with ideas in cognitive psychology which ignored the observed fact that students do not transfer their knowledge across different settings. Brown *et al.* (1989, p. 32) say 'Situations might be said to co-produce knowledge through activity. Learning and cognition, it is now argued, are fundamentally situated.'

An extreme interpretation of this view would suggest that formal instruction was inappropriate, with the best forms of learning taking place 'on the job'. Certainly, all learning is influenced by the environment in which it occurs and instructors need to be reminded that contexts make up an important part of the learning experience. The extreme version of constructivism implies that there is no evidence that an objective reality exists and knowledge is constructed by learners in interactions between themselves and the environment.

These brief descriptions of ideas which have influenced the development of instructional design will provide a framework for our later accounts of the process of constructing science courses. Certainly the production of a structured correspondence text with particular design features, complemented by learning materials based on other media (which will be discussed in Part B), is a tribute to what can be achieved by the application of rational curricular planning and instructional technology (see e.g. Morgan, 1993, for a critique).

In this section we have considered the ideas of a number of influential figures that have significantly affected the design of instruction. One obvious feature is the very different perspectives on learning that they bring to instructional design. The following quotation refers to two of these perspectives.

> To the followers of Ausubel and Gagné the metaphor for knowledge creation is the production line. Assemble the necessary components and the product will appear . . . To the disciples of Piaget knowledge is an exotic bloom which has its own proper season.
> (Davies and Nisbet, 1981, p. 128)

It is the constructivist perspective on learning that is currently dominant in science education and therefore we will examine this further in the next section. Recent work on how students learn, and especially on how they learn science, has forced a reexamination of the effectiveness of instructional practices, in general as well as in science education, and of instructional practices associated with distance learning in particular. These too will be discussed in the next section.

3. CONSTRUCTIVISM AND INSTRUCTIONAL DESIGN IN LEARNING SCIENCE

> To understand text or spoken language one must relate it to schemata for understanding the world. But the goal of science teaching is imparting the new schemata for understanding, schemata not yet in the student's repertoire so how is the student to understand the texts and lessons that impart the new information?
> (Carey, 1986, p. 1123)

Phenomena discovered by both cognitive scientists and educational researchers demonstrate that students often do not understand what they have been taught in science. There is an extensive literature which indicates that children develop ideas about natural phenomena before they are taught science in school. These notions frequently conflict with accepted views in school science and have been variously called misconceptions, pre-conceptions, prior conceptions, alternative conceptions, intuitive conceptions or alternative frameworks (e.g. Driver and Easley, 1978; Helm and Novak, 1983; McDermott, 1984; Linn and Songer, 1988). Science learning is now seen essentially as a process of conceptual change (Hewson and Hewson, 1983), requiring learners to change their prior conceptions to those of scientists, and the way in which such change occurs has been the subject of many studies.

Many misconceptions commonly persist after students have had relevant school or college instruction and to some extent document the failure of much science instruction. One school of thought believes these alternative views are consistent over time and form an alternative theory of science, rather than representing a bag of tricks which learners can call on haphazardly. Di Sessa has the following views:

> the most fundamental problem is the simple fact that students come to physics classes with no theory at all but instead are used to dealing with the world on a catch as catch can basis where it is quite fair to change tactics whenever the problem is minutely varied . . . students who can articulately espouse *any* systematic view of the physical world would be far better prepared for physics courses than those who can be coaxed into reciting the right words yet behave as if every new problem were an occasion to invent another explanation.
>
> (Di Sessa, 1988, p. 61)

One aim of science instruction is to bring students to an appreciation that there are certain scientific principles which have been described as 'abstract general rules about the operation of the world that subsume a larger number of related events' (Linn *et al.*, 1991, p. 7). If students' intuitive views are sometimes unproductive they and their instructors have a problem. Several researchers (Champagne *et al.*, 1980; Clement, 1982; Minstrell, 1982) have suggested it might be useful for instructors to discuss students' naïve beliefs with them, carefully pointing out what is wrong with these ideas and how they differ from accepted scientific ideas. Minstrell describes an intensive demonstration discussion technique which is one way of conducting these interviews. Such discussions may help learners to develop their problem-solving skills, as well as advancing their concept development.

The most popular view on the process by which science is learnt is that of the 'constructivists' mentioned in Section 2. An extensive amount of work on developing scientific concepts, particularly in children, has been done by Driver and others (Driver, 1984; Driver, Guesné and Tiberghien, 1985). They start from the premise that children construct mental models of how the world behaves. One of the problems for science educators is that while these models evolve as the children gain more experience with physical phenomena in the real world, they often conflict with the 'correct scientific' models to which teachers are struggling to provide access. There is considerable evidence that conventional teaching at school and college at all levels is largely unsuccessful at changing the prior conceptions developed by children. In certain domains such as mechanics, informal theories can persist even in university students (Viennot, 1979). It is very important to stress that adults hold these faulty prior conceptions too and that all areas of science have documented examples of such misconceptions. For example, Brumby (1984) found that students of biology assign adaptive characteristics to changes in the individual, rather than to changes to the species via natural selection among individuals. Lewis (1990) found subjects with PhDs in chemistry who used observational data to explain phenomena despite the fact that these observational data appeared at odds with their scientific ideas. This kind of behaviour was noticed in relation to their consideration of questions like 'what would be the best material in which to wrap a drink in order to keep it cold?'

What lessons are there in this for instructors? It would not be sensible to discourage students from forming their own conceptions, even if these turn out to be different from the accepted views of scientists. Such an approach can leave students with the impression that their own intellectual efforts are basically inadequate and they should just learn the ideas of others by rote. This attitude promotes the dangerous idea that science is just a bunch of facts! However students can cling to their own ideas (see e.g. Burbules and Linn, 1988) and shape evidence to defend ideas they have developed (Linn and Pulos, 1988) even in the face of reasoned argument to the contrary. A way forward here is to work first with the productive ideas that students have, rather than to confront them immediately with the fact that some of their ideas are erroneous. This view that making conflict explicit may not be the most useful technique has been commented on particularly in relation to science at the primary level (see e.g. Harlen, 1993; Scanlon *et al.*, 1994) and secondary level (see Solomon, 1989).

Students' understanding of the language used by scientists is also relevant here. Many apparently ordinary words in fact have very precise meanings in science, meanings that may even be in conflict with colloquial usage. Science instructors trying to foster conceptual change in their students may also have to overcome a language barrier. Examples occur in every scientific discipline; obvious ones include:

- *mass and weight*, which novice physics students tend to use interchangeably; also in everyday usage 'massive' just means 'big'.
- *acceleration*, which to non-scientists always implies an increase in speed; the concepts of negative acceleration, and of acceleration at constant speed (i.e. in a curve) are quite alien to many people's intuitive understanding.
- *energy*, which in common parlance usually equates to 'fuel'; if people believe you can 'save energy' by turning down the thermostat, then it is difficult to get them to appreciate the universal truth of the law of conservation of energy.
- *fitness*, which has quite different meanings in biology and on the sports field.

For an instructional designer there is a problem with how to use these constructivist ideas. The case for the importance of educators taking seriously the prior conceptions that learners have is well established and has links with previous ideas in instructional design such as Ausubel's. However there are relatively few documented instances of instruction designed along constructivist principles bringing a demonstrable improvement in performance. One successful example was the 'Conceptual Change in Science' project (Scanlon *et al.*, 1993b) – an attempt to improve a group of children's understanding of some basic concepts in mechanics. It is significant that this example is one involving computer-mediated instruction because there have been more attempts to examine the consequences of constructivism in instructional design related to these type of learning environments (see e.g. Jonassen *et al.*, 1993).

Driver (quoted in Scanlon *et al.*, 1993c) has written about the stages necessary in a teaching sequence to promote conceptual change which she identifies as: orientation, elicitation, restructuring, application and review. By orientation she means some way in which the learner's interest in the topic was aroused. By elicitation, she means some way in which students become aware of their prior conception. Some particular ideas for the restructuring phase which arose from her work designing, trialling and evaluating teaching strategies were:

- broadening the range of application of a conception;
- differentiation of a conception;
- building experiential bridges to a new conception;
- unpacking a conceptual programme;
- the importing of a different model or analogy;
- the progressive shaping of an analogy;
- the progressive shaping of a conception;
- the construction of an alternative conception.

There is a developing tradition in working in this constructivist way. However it is unfortunately true that so far these ideas have had more impact at the school rather than the university level. An exception is the work of Brown and Clement (1987) with college students, highlighting the importance of thought experiments in the restructuring phase mentioned above. The most hopeful aspect is to concentrate on the elicitation phase mentioned by Driver. One clear legacy of the constructivist approach is the imperative of finding out what students already know before attempting to design instruction for them.

Some theorists have attempted to go further with ideas about when constructivist approaches to instructional design are most appropriate. Jonassen *et al.* (1993) takes the view that a constructivist approach is particularly appropriate for advanced knowledge acquisition. Jonassen's diagram (Figure 3.2) shows a continuum of phases of knowledge acquisition, a natural consequence of which is that different types of learning should require different approaches.

We believe that the initial knowledge acquisition phase is better served by instructional techniques that are based on classical instructional design techniques. Classical instructional design is predicated on predetermined learning outcomes, constrained and sequential instructional interactions and criterion referenced evaluation. We believe that constructivistic learning environments may be used during the latter stages of initial knowledge acquisition, and that they represent rich and meaningful environments for initial knowledge learners. However, constructivistic environments are more reliably and consistently applied to support the advanced knowledge acquisition phase. It is at this phase that misconceptions . . . are most likely to result from instruction [based on] classically instructionally designed materials that oversimplifies and pre-package knowledge.

(Jonassen et al., 1993, p. 232)

Jonassen, having established a need for a constructivist approach to instructional design in higher education with more advanced learners, describes some key features of the approach: active learning for knowledge construction, social negotiation of meaning, and contextualization of learning. He reiterates that 'constructivists assume that learners construct knowledge by interpreting perceptual experiences in terms of prior knowledge, current mental structures and existing beliefs' (p. 233) and cites evidence such as that put forward by Gardiner (1989) on the generation effect, i.e. that learners remember material better if they generate it (rather than have it presented to them). The other key planks of Jonassen's manifesto relate to social constructivism – highlighting the social negotiation of meaning has led to a new emphasis on the importance of collaboration in learning. Dialogue is a key pedagogical strategy for constructivist instructional design, as learners can usefully construct their knowledge together (see e.g. Edwards and Mercer, 1987).

Winn takes a similar view to Jonassen in feeling that there is no need to change traditional instructional design practices totally. Winn believes that traditional instructional design is best suited to improving automatic performance; as he puts it, 'Sometimes students need to be taught *how* to do something without having to understand *why*' (Winn, 1993, p. 202, original emphases).

One problem with this analysis is the dearth of exemplars of a constructivist approach attaining demonstrated success at the university level, mainly because there are few examples of such approaches actually being tried. Two other issues remain that are also worthy of mention – the affective and situated aspects of learning. Too often learning is researched as an essentially cognitive phenomenon. Our earlier discussion of social constructivism indicates the folly of ignoring consideration of students' interests and attitudes. When such factors are discussed, it is usually only in relation to students' willingness to study science at all, but the importance of affective issues goes wider and deeper than that. Baird *et al.* (1991a, b) discuss a model of learning science which stresses the importance of balance between cognition and affect for effective teaching and learning. This is central to the ways in which students interpret and work on tasks, and work together on their learning. In Baird *et al.*'s view, which is shared by other researchers (see e.g. Brown, 1988), metacognition has to develop

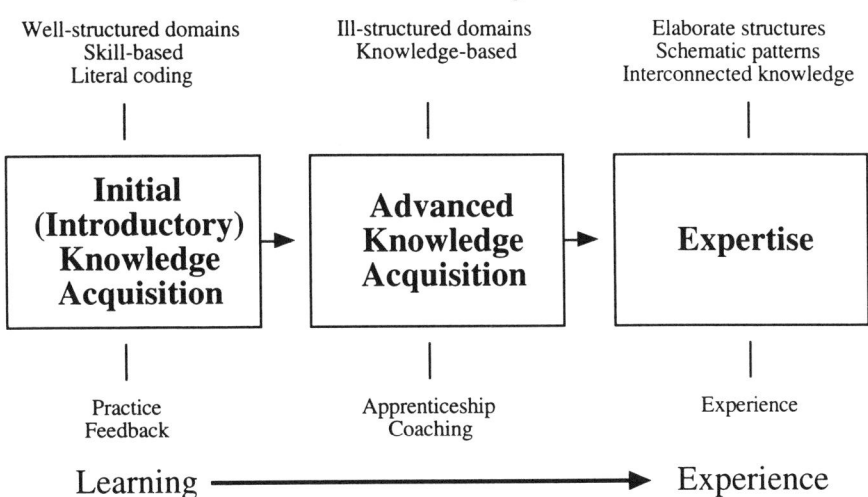

Figure 3.2. A continuum of knowledge acquisition: phases of knowledge acquisition, with associated types of learning, and suggested approaches to learning (redrawing of Figure 1 of Jonassen *et al.*, 1993, p. 232).

in context and depends both on students' knowledge of learning/cognition processes and on their awareness and control of personal learning practices. This emphasis on learning as mediated by critical reflection has become an important influence on the development of instructional design.

One recent study of undergraduate students highlighted the importance of the affective response which was involved in learning. A series of interviews led Entwistle to note that

> Students repeatedly commented that the experience of understanding generally had a feeling tone with it – there was necessarily an emotional response at least where significant understanding had been achieved. The inseparability of cognitive and emotional components of understanding was made very clear in the comments made by the students.
> (Entwistle and Entwistle, 1992, p. 7)

A classic study by Perry (1970), which followed students through their four years at Harvard and Radcliffe Colleges, documented nine stages that students worked through in the development of their thoughts about learning. It is important to realize that students, particularly at university level, will have developed views of the processes of teaching and learning which will alter the way they approach specific learning tasks (see e.g. Säljö, 1989).

In summary, the implication of the constructivist perspective for instructional design in higher education is that learners need opportunities to engage in knowledge construction. According to the social constructivists, the means by which such construction can take place also needs to include opportunities for others to influence the process. Furthermore, context plays an important role in constructing learning situations, and learners need to have a shared context in which they can engage during their joint activity of constructing knowledge. This learning context involves not only the tasks that students engage in, but the others involved in those tasks. Also, the idea that students are agents in their own learning highlights the importance that their past experiences have on their interpretation of the tasks set. In the next section we will discuss how such a perspective influences instructional design for students learning at a distance.

4. INSTRUCTIONAL DESIGN FOR DISTANCE LEARNING

A good description of what social constructivism means for distance education is given by Evans and Nation, as follows:

> It is most important that students are understood as the key agents in their own learning and that both individually and collectively they can (and often do in spite of closed course materials) shape their own learning, not just in the ways they learn, but also what they learn. Thus dialogue should be encouraged not just between teacher and students, but also between students themselves (study groups, tutorial groups, self-help networks, etc.) and between them and those in social contexts within which they live and work. In this respect dialogue should be encouraged through the course materials by providing the students with knowledge, skills, ideas and values which are relevant to their needs and interests, and which they can use actively to understand, manage and change their social worlds through dialogue with their fellows.
>
> (Evans and Nation, 1989, p. 39)

From a constructivist perspective on learning, a vital issue in developing instructional practices for distance teaching is the importance of providing opportunities for dialogue and for collaborative working. It is often assumed that distance learning systems provide no opportunities for such dialogue. On the contrary, however, many distance teaching institutions do provide opportunities for students to meet one another in self-help groups, and to attend various classes. In the UKOU system, regular tutorials are arranged, although attendance is not compulsory. Support for dialogue can occur in both formal tutorials and informal study group or self-help sessions. An UKOU student remembers:

> The tutors allowed you to express yourself, never, ever saying, 'That's rubbish' . . . They made you think, 'That's one way of putting it', and then they'd show you another way without putting you down.
>
> (Lunneborg, 1994, p. 99)

One example of how such opportunities for dialogue and collaborative working can be created is provided by the residential school element which forms a mandatory part of the UKOU's science learning provision (discussed further in Chapter 10). As the following student comments make clear, such schools play a crucial and unique role in giving remote learners access to one another and to tutors:

> I did four summer schools . . . Everybody says there's something incredible about the OU system of summer schools and they're absolutely right. You get there 120 people at a particular week at a particular place. Split them up alphabetically into groups of 10 or 12, stick them into a room with a tutor, and by the end of the week you still have a bunch of 12 total strangers but who are a cohesive group of people who have worked together like a dream.
>
> (Lunneborg, 1994, p. 30)

> The easiest, the most enjoyable, was summer school. I craved someone who wanted to listen to me going on about whatever subject I happened to be studying . . . So it was a release of all the pent-up desire to talk about the course.
>
> (Lunneborg, 1994, p. 88)

However, while these ways of providing the opportunity for dialogue exist in distance teaching systems there is also a need to consider what this perspective on learning means for the construction of other features of the complete distance learning package. Social constructivists stress the importance of others in the learning process and situated cognition stresses the importance of context in learning. The importance of a shared social context for learning by conventional means has led to the perception of the school and classroom (or university and lecture room) culture as vitally important. In conventional classrooms, this shared context is created by the subtle mix of expectations and experience of study. For distance learning students to have meaningful dialogue which leads to learning, this shared context needs somehow to be created. For distance students, the classroom is replaced by the features of the correspondence text, the tutorial session or the assignment practices. The way in which these different components of the distance learning package are structured is often the consequence of traditional instructional design, with much attention paid to the setting of goals or objectives for students, consideration of the most effective sequencing of instruction, and analysis of the role that different media can play in student learning. However, in distance learning institutions one of the most important aspects of the traditional instructional design features which have been developed is quite separate from what may have been their original intent – that is, they help to create the shared context. We would therefore contend that remote students can be helped by coming to share an expectation of what they need to do to study a distance teaching text. Some of instructional design features such as objectives become part of the context that students use to share their learning experiences.

Instructional design in distance learning is often criticized for assuming that the logical structuring of texts and the inclusion of such features as activities in text in fact controls student behaviour, the argument being that such control is neither desirable nor easily reconcilable with constructivist principles. However there is strong evidence that distance learning students in fact use such devices to suit their own purposes. Lockwood (1992) reviews both the attitudes of students towards these devices and those of the writers of self-instructional materials. Among the reasons given for using activities in their teaching, his distance teachers cite fostering learner independence as a major goal, and creating enthusiasm and interest as another. There is little unrealistic expectation of control of the way students work through the material.

Later, we give other examples of such traditional instructional design features for distance learning, including self-assessment activities and continuous assessment practices.

5. CONCLUSION

Instructional design is based on the assumption that learning depends in some predictable way on the instruction that a learner undergoes. There is body of opinion that the application of certain instructional principles can lead to more efficient learning. Recent developments in the study of how students learn, in particular the views of the constructivists, have given a severe challenge to the claims of instructional designers. For distance teachers, the interpretation of learning as being best described as social constructivism has particular consequences. In this chapter we have argued that it is possible to create a shared social context for distance learning, and opportunities for dialogue between remote learners. As a result of this interpretation of learning, distance teachers have also experienced a new impetus into investigating the potential of new teaching methods such as hypermedia or multiple media environments which cater for different study strategies, and exploring new ways of harnessing the potential of group work.

In Part B of this book we describe how distance learning packages are prepared, both in terms of particular design features such as objectives, self-assessment questions and computer-marked assignments, and in terms of how appropriate media are selected for the construction of multiple media packages. Students can use the shared context of such features as objectives, and self-assessment questions to engage in dialogue which will allow them to construct their knowledge. Distance teachers use the language of instructional design to communicate with each other to design appropriate learning experiences. We will discuss how their use of multiple media can be brought together to produce effective science teaching at a distance.

PART B:

Course Design

4

USING MULTIPLE MEDIA

Can distance education exist without the use of media?
(Sauve, 1993, p. 306)

1. INTRODUCTION

In this part of the book we describe the design, development and production of complete distance learning courses in science and their components. In this first chapter we provide an overview of the methods of production of such courses, focusing on two particular aspects – the course team approach and the multiple media approach. In the chapters that follow we illustrate this overview with descriptions of the design decisions made in particular courses and exemplify the particular role that each medium can play in teaching science. Later, in Chapter 12, we discuss how whole courses can be assembled from different components. Of course, selection of media is only one of the decisions to be made. The course design process includes decisions about the course structure as a whole, the specification of aims and objectives, the content and skills development to be included, the teaching strategies to be used, media choice, assessment strategy and evaluation.

The starting point for such design choices is the instructor's view of the teaching/learning process in science, which we described in Chapters 2 and 3. The choice of conventional teaching methods, whether lectures, set readings, exercises and problems, seminars, tutorials, field trips or practical sessions, contains assumptions about what will help students to learn. These different methods each support a number of learning outcomes to a different extent. Presentation of new ideas may be most efficiently done by lecturing, practice in skills by exercises and problems; on field trips and during practicals students often become aware of the pleasures of discovery. In this way a combination of methods provides the appropriate range of educational experience for the learner. For the designer of distance learning courses the particular challenge is to harness the potential of different media so that the learning experiences offered to the distance student at least equal or, even better, surpass those available to conventional students. Romiszowski (1988, p. 8) defines media as 'the carriers of messages, from some transmitting source . . . to the receiver of the message (which in our case is the learner)'. This quotation is somewhat reminiscent of the hydraulic view of education, with the learner as an empty vessel to be filled with knowledge, and conflicts with contemporary views of the student and teacher as partners in learning. Self (1985, p. 22) provides a more useful approach and defines an educational medium as 'a device through which absent teachers are said to communicate with pupils', using an analogy with a spiritual medium – a person through whom spirits are said to communicate with the material world. So in distance education the use of multiple media is required to serve two distinct purposes. First, as in conventional higher education, there is an expectation that the students will be provided with a range of experiences in order to help them learn. Secondly, different media are used to facilitate communication between distance teachers and students.

Take the example of the lecture, which is used as the main teaching method in much conventional higher education. It is efficient in that one lecturer deals with many students, but inefficient because many lecturers in different institutions are presenting the same basic course in mechanics or electricity or enzyme kinetics. It is both effective in that students can be stimulated and inspired, and ineffective in that it cannot take much account of the individual backgrounds that students bring to their learning. The lecturer has to make decisions about entry behaviour and hence about the suitable pace to introduce new concepts. (The approach often adopted is to decide on the level of an average student, make conservative assumptions about their prerequisite knowledge and aim the bulk of the lecture at them.)

What replaces this delivery method for distance students? The medium of choice in most distance learning systems has been text, as we mentioned in Chapter 2, although television has been widely used in many countries to broaden access to courses. The reasons for the predominance of text will be discussed in Chapter 6. However, few distance learning systems rely on a single medium. Most distance teaching universities combine learning from a number of media with some personal contact involving perhaps face-to-face or telephone tutorials or correspondence tuition. One of the first decisions made by the designer of a distance teaching course is the selection and combination of a range of media.

2. MEDIA SELECTION

There is research evidence on the effectiveness of the use of different media to help course designers to choose media appropriately. For example, Olsen and Bruner (1974) contend that content can satisfactorily be presented by any medium, but that different media are better suited to the development of different mental skills. Salomon's (1979) work goes further and argues that certain ways of thinking about subject matter can best be represented by one medium rather than others. This is because media vary in the combinations of symbol systems they employ. Salomon classifies these symbol systems as digital, analogic and iconic. As Bates (1988a) points out, different information is conveyed by a written account, a televised account, a computer simulation and a radio programme giving an account of the same experiment. Different subject areas have different requirements for the representation of knowledge, so one part of the task of media selection is to consider the particular requirements of the discipline being taught.

There are also more general messages to be considered in media selection. Laurillard (1993) offers a classification of educational media. She considers the learning process as a sort of conversation (as do other contemporary writers like Ramsden (1992), using a view first encapsulated in Pask's conversation theory). As a result her classification of media is based on a conversational framework which identifies the activities necessary to complete the learning process. The learning process she asserts 'must be constituted as a dialogue between teacher and student, operating at the level of descriptions of actions in the world, recognising the second-order characteristic of academic knowledge' (p. 94). Thus her classification system is based on the type of interaction between instructor and student that is made possible by the use of a particular medium. As shown in Figure 4.1, she classifies educational media as 'discursive', 'adaptive', 'interactive' or 'reflective'.

Laurillard has used these catgories to engage in a comparison of the particular features of each medium in use in higher education and comes to the conclusion that only two media – tutorial-simulations and tutoring systems (see Chapter 8) – can claim to address the entire learning process. Her

---

Discursive: both teacher's and students' conceptions are accessible to the other and both topic and task goals can be negotiable; students must be able to act on, generate and receive feedback on descriptions appropriate to the topic goal; the teacher must be able to reflect on student's actions and descriptions and adjust their own descriptions to be more meaningful to the student.

Adaptive (by teacher): the teacher can use the relationship between their own and the student's conception to determine the task goals for the continuing dialogue, in the light of the topic goals and previous interactions.

Interactive (at the level of actions): the students can act to achieve the task goal; they should receive meaningful intrinsic feedback on their actions that relate to the nature of the task goal; something in the 'world' must change observably as a result of their actions.

Reflective: teachers must support the process by which students link the feedback on their actions to the topic goal, i.e. link experience to descriptions of experience; the pace of the learning process must be controllable by the students, so that they can take the time needed for reflection when it is appropriate.

Figure 4.1. Classification of educational media (taken from Laurillard, 1993).

conclusion from this analysis is not that instructors need confine themselves to these two media but that educators should use such categorizations to become aware both of the deficiencies of a particular medium and of the potential for construction of learning packages that combine complementary features.

> Stand-alone media-based packages will never be sufficient, because none of the media can adequately support the discursive activities that are essential for academic learning. However, the media comparison shows how to integrate a range of media in order to best exploit the strengths of each. Improvements in university teaching are less likely to come from 'multimedia' than from 'multiple media'.
>
> (Laurillard, 1993, p. 176)

Most other approaches to classifying media have been a little more pragmatic and various writers (e.g. Rowntree 1974; Hodgson, 1984) arguing from such perspectives come to the same conclusion as Laurillard herself that the best approach is more likely to require an integrated combination of several media rather than a single one, and that several media combinations will provide adequate alternatives. Schramm (1977) expresses this pragmatic view of media selection quite well:

> For any given task in any given situation, one medium is likely to perform better than other. However, the rather more important conclusion is that most media of instruction have a wider spectrum of usefulness than is sometimes appreciated. If the medium that seems ideal for a specific purpose is not available . . . it becomes therefore important to consider how second- and third-'best' media choices can be most effectively used.
>
> (Schramm, 1977, p. 268)

Choosing the best medium for a particular piece of instruction is more likely to depend on logistics than on pedagogy. We illustrate this for different media in the succeeding chapters.

3. LOGISTICS

There is no point in engaging in a sophisticated educational analysis of a teaching problem if the solution is impracticable. Countries and institutions have developed different solutions to the delivery of distance learning due to such factors as the availability of broadcasting, the reliability of the postal system, the ownership of televisions, and the geographical distribution of the population (which together with the availability of suitably trained teachers would influence whether face-to-face tutorials were ever feasible). Part of this logistic analysis needs to consider comparative development costs. Costing distance education is not easy. Assigning costs is not simple and deciding what the effects of the teaching are is equally difficult. Further, if the cost-effectiveness of a particular media mix is being assessed, some comparative cost data need to be assembled (a point first made by Fielden and Pearson (1978)). The most recent interesting work has been done by Rumble (1987), following on from the work of Perraton (1982a), and adopting what has become the conventional view of costing. Hawkridge (1991) has recently provided a discussion of the problems associated with costing media and assessing effectiveness which concludes that such an analysis is extraordinarily difficult.

Laurillard, who has also looked at costing, writes

> A multiple media course at the Open University, for example, might distribute student study time spent on print, video, computer, tutorial, essay in the proportions 10:1:2:1:3. The academic development time spent for that distribution of teaching would be equivalent to the proportions 60:10:10:1:1, assuming they (academics) get no support for writing the text, or for designing the tutorial and the essay question, but would get substantial support for the development of the video and computer material, where their input is academic design only. Thus print appears to make most efficient use of academic time in terms of student study time generated, and setting essays is most efficient. If efficiency is measured in terms of student study time for learning gain, however, then it is possible that the computer or video could perform better.
>
> (Laurillard, 1993, p. 206)

Bates estimates a ratio of 10:1:10–20 for production:delivery:support costs for distance teaching universities. Notice that he puts the support costs highest (Bates, 1991, p. 12). This pragmatic approach is not intended to deny the particular characteristics of any one medium and we will be using the succeeding chapters to illustrate their individual potential. For example, print has specific advantages which have been obvious to correspondence colleges for many years. However, print also has some deficiencies as an instructional medium. It shares many of the weaknesses of the lecture, in that it is usually neither interactive or adaptive. Some of these weaknesses can be removed by incorporating such design features (called access devices) as learning objectives, in-text questions and activities. More will be said of this in Chapter 6. Similarly we will analyse audiovisual media and computer mediated learning in Chapters 7 and 8 respectively.

4. THE NEED FOR TEAM WORK IN COURSE PRODUCTION

The design of a multiple media course is not a simple task. One of the particular contributions of the UKOU to the development of these types of courses is the course team approach. Each course produced is the result of work done by a team trusted by the university with the total responsibility for the academic content of the course and the way in which it will be taught. Starting from the approval of an outline proposal that a course on some specific topic will be prepared, the team has to map out in detail the material to be covered and decide how the teaching methods at its disposal can best be used for the particular target student body. Members of the course team include subject matter experts from the relevant Faculty, media production experts (such as graphic or software designers or computer programmers), the BBC producers who will be responsible for the audiovisual components and educational technologists. Part of the reason for the team approach is that few individuals have the mix of skills necessary for the solo production of an entire course. However, each member has to be more than just an expert in a particular mode of educational communication, and is expected to interact with the other team members and to make a contribution to the course objectives and content. One of the benefits of this system is that course team members learn from each other. An early description of a course team's operation described the tasks that needed to be completed in the 'systems approach' language common at the time as:

> to specify educational objectives; to break down the students' task into component elements, or successive stages; to identify the learning processes involved; to consider the appropriateness of the teaching methods available in relation to each stage; to combine the methods so they make an integrated whole; and finally to provide for feedback validation and assessment.
>
> (Mackenzie et al., 1975, p. 339)

UKOU courses are organized in terms of units of work, each involving typically a week or fortnight's study time for the student and once the course team as a whole has decided the overall course plan, the responsibility for each unit is placed in the hands of a single author (or sometimes a small group). Each component, whether correspondence text or audiovisual material has to be scrutinized in draft form, usually several times, until the course team approves the final version for delivery to students.

The UKOU's first Vice-Chancellor, Walter Perry, regarded the course team concept as the institution's 'most important single contribution to teaching practice at tertiary level' (p. 91), and believed that the course team provides the 'drive for innovation that is the essential part of maintaining the Open University in good health'. The UKOU continues to use this team approach, although some difficulties have been reported (see e.g. Stringer, 1980) and it has been less popular in other distance teaching institutions. It will be discussed further in Chapter 12.

Of course, a course team works under one severe disadvantage compared with conventional teachers – lack of contact with the students who are the recipients of the learning package. This means that closing the loop of the feedback cycle is a vital part of the job of educational technologists on course teams – a topic which is addressed in Chapter 14.

5. MULTIPLE MEDIA APPROACHES TO AID SCIENCE LEARNING

It is difficult to overemphasize the scepticism that greeted initial reports that the UKOU was to teach science to remote students without prior qualifications in the subject. The three main reasons why teaching science at a distance to students without previous educational qualifications in science is difficult – the hierarchical nature of science knowledge, the necessity to have sufficient background mathematics and the requirement to find a substitute for long hours in the laboratory – were discussed in Chapter 2. The intertwined nature of the subject matter, and the fact that an understanding of some concepts require mastery of prerequisite knowledge, do make severe demands on course design for distance teaching. However, the multiple media approach has particular benefits for science learning. As well as the obvious requirement of finding alternative media to replace some aspects of the experience of a science student in a conventional higher education institution, a number of studies demonstrate that for difficult concepts, presentation in a different medium may aid student learning, and that as we discussed in Chapter 3, different levels of representation can be used by students at different stages in their learning. Bates (1988a) illustrates how this might work for a science topic as follows:

> to comprehend fully the concept of heat, one needs to experience it in a number of ways; physically, verbally, numerically, conceptually, symbolically. Thus the concept of heat can be 'experienced' by touch (burning oneself as a child), through words ('that is hot') or numbers (42 degrees Centigrade) associated with the physical experience, by abstract definition ('a form of energy arising from random motion of molecules of bodies'), or through symbolism ('a man dragging himself through a desert'). Each way of 'knowing' about heat is different. Eventually though the learner must integrate all these experiences to understand fully the concept of heat. Different media can provide distinctly different experiences or representations of the same concept and this becomes especially important for students where important concepts or experiences are not likely to be common in their everyday life.
>
> (Bates, 1988a, p. 223)

In the rest of Part B we analyse for each medium its potential for distance teaching in general and for the teaching of science in particular. In each chapter we will be dealing with one type of medium such as text or audiovisual components, or one aspect of course design such as evaluation, or one type of course provision such as tutorial support, and describing some successful attempts to overcome the problems we have outlined in Chapter 2.

5

SKILLS DEVELOPMENT

It could be said that the most valuable elements of a scientific education are those that remain after the facts have been forgotten.
(Screen, 1986, p. 13)

1. CONTENT, PROCESSES AND SKILLS

Science, perhaps more than other disciplines, is often represented as consisting of two distinct but interwoven strands – that of knowledge and that of process. The first is concerned with the collection of scientific concepts and principles – the traditional stuff of old-style textbooks. The second relates to the ways in which scientists actually go about their jobs and the strategies they use in problem-solving. One of the most obvious recent trends in science education, in Britain and elsewhere, has been the move away from content-led curricula towards process-led approaches. This has sparked a lively debate about the role of skills development and process in enabling science learners – at all levels – to attain the ultimate goal that Woolnough has characterized as becoming 'good at doing science'.

> There is no doubt that a scientist needs to possess specific skills, skills of observation, of measurement, of manipulating apparatus, of recording and interpreting data and of communicating with others. But these skills should be seen as having their *raison d'être* in the whole scientific activity, as part of an investigation. Their value as ends in themselves is limited. The prime aim of a scientific education should be that the student has become good at doing science with the ability to perform genuine scientific investigations.
>
> (Woolnough, 1989, p. 118)

The process-led approach can be seen as having grown out of the ideas of discovery learning discussed in Chapter 3. First came curricula focused on practical work and enquiry learning, of which one of the most influential was perhaps the Nuffield science initiative introduced in English secondary schools in the late 1960s. Then during the 1980s, there was a dramatic rise in process-led courses (see, for example, the review by Murphy and Scanlon

(1993) of initiatives for younger children), where the emphasis was, if not directly on practical work, at least on investigative approaches. The advocates of the process approach believed it would help learners to acquire high level skills, which would be transferable across a wide range of applications. By that time, however, the dominant views in science education had moved towards the constructivist perspective, and evidence from empirical studies was demonstrating that most practically-based instruction was failing to allow learners to construct new models that they could integrate with previous knowledge and understanding. These critiques lead to some helpful elaboration of what it meant for students to be 'active' in their learning: 'We see the students essentially active in the learning process in which they are continually enquiring, testing, speculating and building up their own personal constructs of knowledge' (Woolnough and Allsop, 1985, p. 33).

The view of Woolnough and Allsop was that practical and investigative work is important in allowing students to learn how to 'do' science, but that it should also be structured so as to develop their conceptual understanding. A seminal article by Miller and Driver (1987) extended this thinking into a critique of the whole approach of process-led curricula. They, and Jenkins (1987), argued that the idea of a prescribed set of processes adding up to *the* scientific method was fundamentally flawed. They also took issue with the assumption that students could progress in scientific understanding simply by exercising general skills and so discovering knowledge. This of course goes to the heart of the constructivist view of learning, which suggests that conceptual understanding is advanced only when links are made between models and experiences, and not through the acquisition of general rules and techniques unrelated to domains. There is some conflict here with the notion of 'transferable skills', and this is an issue we shall explore in the next section.

None of this is to say that there are not skills, techniques and strategies that science students have to learn and practise. The constructivist view does not deny the importance of process, but argues that it is only in their contextual framework that such processes have scientific meaning and instructional value.

> Without some conceptual understanding the problem to be investigated will have no meaning . . . The carrying out of an investigation does not rely on the ability to explain a phenomenon, only to recognise its significant aspects and to develop an effective plan of action. In other words tacit rather than explicit understanding of the concepts involved can prove adequate. Furthermore as procedures are deployed in the course of the investigation, there are opportunities to amend that plan of action to take into account incomplete conceptualisation or inadequate planning. Problem-solving relies on both the procedures of science and conceptual understanding but it can also act as an important vehicle for their future development.
>
> (Gott and Murphy, 1987, p. 8)

Although based on studies at secondary level, Gott and Murphy's view of the interdependence of conceptual knowledge and process achieved a broad measure of support across many levels of science education. Their terminology for scientific investigations distinguished between 'conceptual understanding', which leads to the identification of the key variables in a problem, science 'processes', which include activities such as observing, communicating, predicting, inferring and interpreting, and scientific 'procedures', which encompass the strategies common in scientific inquiry, such as varying one quantity while keeping others constant. Millar (1989) has adopted a similar stance with respect to practical work, considering Gott and Murphy's categorization of procedural understanding to be too narrowly focused. He distinguished between 'cognitive processes', such as observation and classification, and practical skills, the main difference between the categories being that, in his opinion, practical skills could be taught and cognitive processes could not.

Skills and processes have been subdivided in different ways by different authors. Watts (1991, p. 14) has offered as working definitions: 'a skill is an individual's contribution to an activity, or towards an objective', whereas 'a process is a more general means to an end'. His classification system leads to the following framework:

strategy = sum of processes
process = sum of skills

Nevertheless, as Watts himself acknowledges, it is not always clear at what point skills shade into processes. The ability to describe something clearly is a skill; communicating clear descriptions is a process. In the next section we focus in more detail on the skills that science students are required to master, without concerning ourselves too much with semantic distinctions between skills and processes. However, the mastery aspect is important: to say that a student has acquired a skill is not quite the same as saying they are 'skilled' (in the colloquial sense of the term) at some particular activity. This kind of discrepancy had led in some subject areas to a move away from the terminology of skills towards that of 'competencies', though this trend is not yet so pronounced in science instruction as in other areas. The definition of competence varies with the discipline, but encompasses the idea of theory and practice coming together into action. Skills may thus be thought of as enabling performance and contributing to competence.

2. SKILLS DEVELOPMENT IN SCIENCE

If there is a diversity of opinion as to what exactly constitutes a skill, and whether the application of skills in combination is equivalent to a process, there is even less agreement about the ways in which skills may be classified. In a system especially geared towards instructional design, Romiszowski (1981) has identified four main skills categories: cognitive, psychomotor, reactive, and interactive. Following the ideas of Briggs (1970), he further subdivides each category into 'reproductive' (i.e. algorithmic) skills and 'productive' skills (which involve the application of principles or strategies), and notes that these map well on to Bloom's taxonomy of cognitive objectives listed in Figure 3.1 – knowledge/comprehension/ application being classed as reproductive, and analysis/synthesis/evaluation as productive. Some of Romiszowski's categories are more directly relevant to science instruction than others. For instance, his cognitive category covers skills of decision-making, problem-solving, logical thinking, etc. An example of a reproductive skill in this class would be 'substituting given values into a formula'; examples of productive skills would be 'solving a new problem', or 'devising a proof of a geometric theorem'. Psychomotor skills are also clearly extremely relevant to some aspects of science teaching and learning, especially practical work. Reactive skills (dealing with oneself) and interactive skills (dealing with others) are less obviously covered by conventional scientific training, except perhaps to the extent that they encompass the development of 'scientific attitudes' – curiosity, open-mindedness, healthy scepticism, willingness to consider other points of view – and the ability to work co-operatively within a team. We will return briefly to the skills of communication and group working towards the end of this chapter.

However, the learner (and the instructor too, for that matter) usually starts from the other end, with a list of skills that are acquired individually. If the skills are grouped, this is done first under contextual headings. Thus, for example, a foundation level course in science might include under the heading 'data handling skills' such things as the ability to tabulate data, draw graphs with appropriately scaled and labelled axes, draw 'best-fit' lines, place error bars on graphs and appreciate their significance, etc. It can be argued that proceding from the particular to the general in this kind of way is the most effective from the point of view of instructional design.

If skills are broken down sufficiently far, some are clearly peculiar to science or indeed to some particular discipline of science. The ability to carry out a titration, use a petrological microscope or draw the Lewis structure for a compound are examples of these sorts of rather narrowly focused skills. Other types of skills, such as the formulation of hypotheses and the testing of predictions by experiment, are specific to scientific *methodology*, but may also be applicable in other spheres. Then there are the kinds of skill that are useful in a wide range of contexts – examples include the ability to recognize and record distributions, patterns and relationships, measure and estimate quantities, weigh and interpret evidence, or present a logical argument. Skills that are common to many contexts and contents are sometimes called 'core' or, more commonly, 'transferable' skills. Those who use the latter term are reflecting a conviction that it is indeed possible to develop domain-independent skills. That is not to say that it is possible to teach or to practise skills in any kind of contextual vacuum. As discussed in Section 1, skills have to be attached to both context and content. However, there is a substantial body of opinion that such attachment need only be temporary, and that some skills, once mastered, are remembered longer than the domains with which they were associated when being learned: '. . . it is in the nature of skills that they may be retained and practised long after the detailed content of a course has been forgotten' (Open University, 1991, p. 11).

The quotation heading this chapter is clearly based on such a premise. This kind of argument has been applied both to narrowly defined skills (e.g. graph plotting), and to broadly based skills (e.g problem solving). Glaser, for example, has suggested, that it is possible to involve students with science content in such a way that knowledge and transferable skills develop together. In his view (1984, p. 25), this can be achieved by teaching 'specific knowledge domains in interactive, interrogative ways so that general self regulatory skills are exercised in the course of acquiring domain-related knowledge'.

Any attempt to classify skills, especially those related to a hierarchical knowledge structure such as science, leads one to question whether it is possible (or productive) to devise a 'skills ladder'. Is there in fact any point in ascribing levels to skills? One approach is to label as 'higher order' skills any that subsume more elementary ones, or involve the exercise of a number of different skills simultaneously. This kind of grading exercise is, however, always problematic and frequently inconclusive. Moreover, unless the skills are defined with extreme precision, their intended level of sophistication depends more on context than on their place within some arbitrary hierarchy: a primary school child and an undergraduate science student could both be expected to be able to 'draw conclusions on the basis of observed data'. Figure 5.1, which lists some of the 'attainment targets' of the 1989 version of the national curriculum for schools in England and Wales, illustrates fairly clearly that context and knowledge of the students' prior level of experience are often crucial in interpreting even carefully formulated skills descriptions. Such examples suggest that the ordering and grading of skills can really only be done at the level of the individual lesson, text or course.

A different way forward was adopted by the UKOU Science Faculty in a profile review undertaken to progress, among other issues, skills development in its entire suite of courses. A general framework identifying clusters of skills was drawn up, with the intention that this should serve as a common basis for the more detailed lists that each course team would generate. While there was no suggestion that it would be necessary, or indeed even appropriate, for skills from every cluster to be covered in every course, it was felt that this structured approach would promote good practice and facilitate the interchange of information between academics on different course teams and in different departments in the Faculty. The framework appears as Figure 5.2. It is of course possible to group skills in very many different ways, according to the purpose underlying the classification. In drawing up a list of skills associated with science problem solving (that is with investigative science in its widest sense), Watts includes – alongside categories such as 'observation and visual skills', 'numerical skills' and 'physical and practical skills', all of which map well on to the framework of Figure 5.2 – the interesting cluster of

Imaginative skills: The ability to put oneself into other situations, whether of time, place or person, to visualize other experiences; the ability to discipline the imagination by evidence and experience, to reorder and reshape experiences and images.

(Watts, 1991, p. 41)

LEVEL	STATEMENTS OF ATTAINMENT
	Pupils should:
2	• record findings in charts, drawings and other appropriate forms
3	• formulate hypotheses, for example, 'this ball will bounce higher than that one' • record experimental findings, for example, in tables and bar charts
4	• formulate testable hypotheses • plan an investigation where the plan indicates that the relevant variables have been identified and others controlled • record results by appropriate means such as the construction of simple tables, bar charts, line graphs
6	• prepare a detailed written plan [for a line of enquiry], where the key variables are named and details of the experimental procedure are given • record data in tables and translate it into appropriate graphical forms

Figure 5.1. Extracts from 'Attainment target 1: Exploration of science' of the national curriculum for primary and secondary education in England and Wales (DES, 1989). Levels 2, 3, 4 and 6 refer to the performance that would be expected on average from schoolchildren of ages 7, 9, 11 and 15 respectively. It could be argued that some of the attainments listed here without explanatory examples could equally well describe the skills expected of a postgraduate research student!

Skills of this kind, and the attitudes underlying them, are seldom articulated. They are often hidden under umbrellas such as 'brainstorming' or 'selection of approaches to suit the problem'. But it is in the phase preceding the selection that the real creativity lies – the willingness to ask 'what if . . . ?', the ability to imagine situations from different frames of reference, the courage to entertain apparently absurd (or sometimes simply unfashionable) ideas. These skills are the hardest to foster, but the ones that the best science students can perhaps least afford to ignore.

3. STUDY SKILLS

A skills-based approach to study should not confine itself to developing competencies associated with particular knowledge domains. Indeed, without at least some study skills, learning can hardly even begin. Explicit attention to appropriate study skills should therefore ideally be an integral part of all teaching and learning.

Advice to students on study methods may be offered in many different ways, but Taylor (1984) has pointed out that it normally falls into one of two categories – either it emphasizes *techniques*, or it emphasizes *process*. Examples in the first category include many books of the general 'how to study' variety, which describe skills such as speed reading, methods of note-taking, précis-writing, the construction of study or revision timetables, and so on. There are also books that present particular methods in a generalized way – for instance the one by Lewis (1976) on essay-writing, or those of Buzan (1974, 1993) on so-called 'mind mapping'. However, although study advice focused on techniques has its uses, it is certainly not the whole answer to learners' study difficulties. Such an approach is prescriptive, based on the assumption that there are 'good' study methods, which the learner has only to master in order to be successful. Yet, as Taylor points out, there is no research evidence for a causal relationship between study habits and academic success at university. A 'how to study' programme usually does not make much allowance for differences between individual students, nor does it enhance students' understanding of their own learning processes in a way that might enable them to learn more effectively or more efficiently. It is therefore necessary to look beyond mere technique in seeking to define what one might call 'the skill of learning', and to consider results as well as behaviours. Svensson (1977) has given a description which encompasses both these aspects; he defines a study skill as 'the relation between study activity and the outcome of learning'. According to this kind of process-oriented definition, the development of study skills should bring the learner an awareness of possible study strategies, and an ability to direct their own learning.

Students should certainly be brought early on to an appreciation of the two main categories of instructional and learning strategies. Most 'formal' teaching, and very often also students' expectations of learning, is based on

PRACTICAL SKILLS
such as:
Following instructions and given practical procedures
Planning and carrying out observations, investigations and experiments
Using specific procedures and items of equipment
Assessing accuracy and reliability of measurements

SCIENTIFIC MATHS SKILLS
such as:
Mathematical modelling (including problem formulation)
Problem solving (including the application of numerical and computational methods)
Thinking spatially and visualizing in 3-dimensions
Analysing data (graphical analysis, compounding errors, statistical analysis)
Applying mathematical principles and models in new contexts

DATA HANDLING and RECORDING SKILLS
such as:
Recording data (e.g. constructing tables, etc.)
Displaying data (e.g. drawing graphs, etc.)
Storing and retrieving data (e.g. using computer databases)

SCIENTIFIC METHODOLOGY SKILLS
such as:
Applying scientific principles in new contexts
Devising and using classification systems
Scientific modelling
Reformulating hypotheses and making predictions
Evaluating models and using scientific evidence
Judging the quality of scientific evidence and argument
Dealing with incompleteness, uncertainty and natural variability

COMMUNICATIONS SKILLS
such as:
Presenting information (writing, speaking and demonstrating to a variety of
audiences in an appropriate way)
Arguing cases and presenting different viewpoints
Gathering information (observing, listening and reading effectively
from a variety of sources)

PERSONAL AND ORGANIZATIONAL SKILLS
such as:
Study skills (effective use of given objectives)
Organizing work
Working in pairs or groups at a variety of tasks
Managing time and resources
Identifying personal characteristics and abilities
Developing independence and self-reliance

Figure 5.2. Skills clusters proposed by the Open University Science Faculty (1991), *op. cit.* The various examples given within each category were not intended to be exhaustive, but only to convey the meaning and range of the category heading.

mathemagenic devices: activities (e.g. inserted questions) or signals (such as objectives and advance organizers) imposed by the instructor in order to guide the learner to specific content. Such devices will be discussed in more detail in Chapter 6, where we will argue that they form the foundation of most instructional design for distance education texts. However, there is evidence that students in distance education may not make best use of mathemagenic devices, and that their long-term cognitive gain may not be well served by such strategies (Clyde *et al.*, 1983; Marland *et al.*, 1990). Beudoin (1990) has also highlighted the concern that exclusive reliance on mathemagenic methods creates a dependency in the students, and does not equip them with the tools to fulfil their full learning potential. *Generative* learning strategies, on the other hand, require learners to create their own individual representations (e.g. notes or concept maps) of the material to be understood. Generative methods encourage students to build on their previous knowledge and are thus very much in line with constructivist principles; although the constructivist view was discussed in Chapter 3 principally in relation to conceptual development, there is no reason to suppose it cannot be applied equally well to study methods. Generative learning strategies have been found to appeal particularly to students who have previously lacked workable strategies (Clark, 1990), or who have a strong preference towards visual forms of presentation (Feldsine, 1987; Okebukola and Jegede, 1988). There is also evidence that once generative learning strategies have been well practised, they are especially useful in helping students to master conceptually hierarchical content (Feldsine, 1987; Novak, 1990; Schmid and Telaro, 1990), which of course makes them particularly powerful tools for science students. Gibbs (1981) and Northedge (1990) have produced activity-based books designed on constructivist lines to help students assess their own study habits and develop further strategies appropriate to particular types of study task.

Gibbs (1984) has also analysed various conceptions of learning, and notes that the conceptions held by individual learners may evolve as they come to understand the differences between surface and deep approaches to study.

> Students . . . come to studying with their own orientation which may itself change over time . . . Teaching students to learn must be primarily concerned with students' purposes and their understanding of the nature of the tasks facing them. Such purposes and understanding are not amenable to advice and training. They are deep-rooted characteristics of the developing individual. Conceptions of learning inherent in some study techniques (such as speed reading or patterned note-taking) cannot simply be taken on board and substituted for existing conceptions . . . [Students must] examine their own study habits and beliefs and evaluate their own effectiveness in achieving their own purposes.
>
> (Gibbs, 1984, pp. 281–2)

The great disadvantage of any programme devoted primarily to the development of study skills is that it is divorced from a genuine learning context.

Generalized study advice can take little account of the differences between fields of academic study; yet there is a world of difference between the types of essay expected of physics and history undergraduates. Taylor points to a considerable body of evidence showing that context and perception of the nature of the work to be done are of vital importance in defining any study task:

> In any study task – reading a book, taking notes from a lecture, or writing an essay – it is crucial for the student to know how the task is perceived by the teacher in order to know how to go about completing it.
>
> (Taylor, 1984, p. 272)

The most useful and effective study advice is therefore that given in context, either as an integral part of the course material, or in association with some particular assignment the student is required to complete. However, there are dangers even with this approach. An attempt by one UKOU technology course to weave the development of study and writing skills through the whole fabric of the course was judged only a partial success (Kirkup, 1984): the main dangers were discovered to be the fragmentation of the teaching material, the tension between learning new techniques alongside new content, and the resistance of students to certain unfamiliar techniques. The third of these difficulties echoes the warning from Gibbs, quoted above, that learners cannot substitute one conception of suitable study behaviour for another simply on the basis of 'expert' advice. Such caveats need to be taken into account, but should not absolve instructors from the responsibility of ensuring that their learning packages include explicit material covering ways in which those packages might be approached; in order to construct such material, instructors need to be able to imagine how the package might be perceived by a variety of students, and to suggest sensible study patterns to meet their different needs.

4. EMBEDDING SKILLS DEVELOPMENT INTO INSTRUCTIONAL DESIGN

4.1. An integrated package

Recent trends in instructional design have increased the emphasis on skills development, making it not just an implicit consequence of any study in a particular field, but an explicitly taught and assessed dimension of coursework. Skills teaching therefore has to be incorporated into course design from the outset, and reflected in the objectives that students are expected to achieve. Instructors also have to recognize during the initial phase of course design that skills development can be time consuming, and that some reduction in the knowledge-based content of courses has to be made to allow time for the practice and consolidation of skills.

The integration of a skills-based element into core course material has some very obvious benefits for the learner. For one thing, it promotes an active rather than a passive approach to learning. Secondly, it can give students a better appreciation of what it is that they are trying to achieve: although most objectives, especially at university level, tend to specify attainment targets in terms of knowledge and understanding, objectives framed in behavioural terms are often more useful to students, increasing their awareness both of what is being demanded of them and of how they should be able to demonstrate their competence. Figure 5.3 illustrates this point, and we will return to issues surrounding the use of objectives in Chapter 6. Thirdly, as the emphasis on skills – especially those perceived as transferable between contexts – becomes increasingly important in vocational training, so the skills development aspect of more academic courses can have a significant effect on their validation and attractiveness to students.

However, a focus on skills development is not only of use to the learner. There are very practical benefits for course authors too. Attention to the skills dimension can be extremely helpful in suggesting ways of structuring the material at the level of the individual 'building blocks' of the course. It is also useful, as illustrated in Figure 5.4, in helping authors to identify desirable outcomes of the course in terms of student performance, rather than simply in terms of the assimilation of factual information. A skills audit across whole sections of a course makes authors more aware of the demands that the course work is placing upon students, possibly revealing points at which those demands are excessive or unfair. Identification of such critical points is the first step to providing targeted help for students, and hence increasing retention and pass rates.

A skills audit is also very useful in designing assessment strategies and tests, as illustrated in Figure 5.5. In general, authors should be wary of taught skills that cannot easily be assessed!

4.2. Skills development for the remote learner

A feature of distance education is that study skills acquire a greater importance than in the mainstream sector and the remote learner *may* require different, or additional, study skills from a conventional student. For example, students in distance education may need to develop particular skills associated with learning from broadcast television or radio. However, the fostering of the 'independent learner' and 'knowledgeable communicator' is increasingly being seen as an important facet throughout institutions of higher education, so in some sense the distance learning sector may be showing the way.

As already noted in the previous subsection, if skills development is to be fully integrated into coursework, it really needs to be built in right from the initial planning phase. Among the issues to be resolved at the design stage are the points at which particular skills will be introduced, and even more importantly, where and how often they will be *practised* through worked examples, exercises

1982 version

Now that you have completed this Unit, you should be able to:

- Add two vectors by addition of components, and by triangle addition (*ITQs 28, 29, 36 and 37*)
- Solve projectile problems at the level of the worked example in this Unit (*ITQ 32; SAQs 5-9*)
- Recall the magnitude and direction of the centripetal acceleration of a body moving in uniform circular motion, and apply the equation $a = v^2/r = \omega^2 r$ to simple problems (*SAQs 10 and 11; ITQ 46*)
- Give definitions of angular displacement, angular speed (ω), and period (T), for a point moving in uniform circular motion
- Recall and know how to use the relationships $\omega = 2\pi/T$ and $v = \omega r$ for uniform circular motion (*ITQs 42-44*)

1994 version

Now that you have completed this Unit, you should be able to:

- Relate the components of a vector to its direction and magnitude (*ITQs 3, 11 and 14; SAQs 1 and 2*)
- Resolve a given force into components and add forces together (*SAQs 3 and 4*)
- Solve problems involving the motion of particles on inclined planes, taking into account forces such as gravity, the reaction force and friction (*Problems 1 and 2*)
- Solve problems involving the motion of projectiles in a vertical plane, taking into account gravity but neglecting air resistance (*SAQ 5; Problems 3 and 4; Tape Example*)
- Recall the equations $\omega = 2\pi/T$, $v = \omega r$ and $a = r\omega^2$, relating the angular speed, orbital period, radius, speed and acceleration of a particle in uniform circular motion (*ITQ 20; SAQs 6 and 7*)
- Solve problems involving the forces acting on particles in uniform circular motion, including those in which the centripetal force is provided by gravity (*SAQs 8-11*)

Figure 5.3. Examples of objectives based on a second level UKOU physics course (*Discovering Physics*, Open University, 1982, Unit 2, p. 57) and its rewritten version (*Discovering Physics*, Open University, 1994, Unit 2, p. 38). Both sets of objectives cover essentially the same material, but the later version dwells less on the recall and formal aspects, and is more explicit as to how students are expected to demonstrate their mastery of the concepts. In both cases, the objectives are related to ITQs (in-text questions), SAQs (self-assessment questions) or other teaching devices such as worked examples, to give students an appreciation of the kind of problem they might be required to solve.

for the student and assessed assignments. Figure 5.6 gives a broad overview of the planning process for one course, and exemplifies the role the different media may play in the skills package. The particular strengths of each medium in promoting certain types of skill will be further explored in subsequent chapters, and we will return in Chapter 12 to the way in which the course referred to in Figure 5.6 was put together and its various components integrated.

[students are expected to develop]

- an *understanding* of the ways in which data [can] be presented, and [of] the *interpretation* of such data.

- [the ability] to *work confidently with unstructured information.*

- *the ability to access* the primary and biochemical literature and to efficiently *abstract* information from it.

- the ability to *organize* information abstracted from the primary literature and to construct a framework around which to write a review.

- [the ability] to *write independently* a critically argued review.

Figure 5.4. List of some of the skills specifically developed in a fourth level UKOU course *Exploring the Chemistry of a Neurotransmitter* (Open University, 1991, *op. cit.*, Appendix 4.1, original emphases). This course, at the highest level for undergraduate studies at the UKOU, is designed to foster the kinds of competencies that might be expected of students wishing to embark on postgraduate research work.

Overview of skills tested by the first set of assignments

CMA 41 and TMA 01 have been designed to test not only your knowledge of Block 1, but also the skills you have acquired whilst studying it. The nature of the different parts of the assignments is described below, and forms a progression towards higher order skills, so you should complete the CMA first, and then the TMA.

CMA 41

Part I (50% of marks) largely tests your *recall and understanding* of some key points.

Part II (25% of marks) requires you to read an extract from a research paper, and tests your *recall and understanding* of it.

Part III (25% of marks) also involves reading a research paper, and requires you to *apply* your knowledge of sedimentary environments to *interpret* assemblages of facies.

TMA 01

Part I (35% of marks) involves the *description and interpretation* of two 'unknown' thin sections and requires you to *apply* your petrographic knowledge.

Part II (65% of marks) requires you to study graphic logs and facies descriptions, and make *interpretations* based on these. The last question also requires you to *synthesise* your interpretations in order to discuss the lateral variation within two sedimentary cycles.

Figure 5.5. Part of a skills audit from a third level UKOU course *Sedimentary Processes and Basin Analysis*, which prefaces the first assignment booklet for the course (Open University, 1991, *op. cit.*, Appendix 2.4, original emphases).

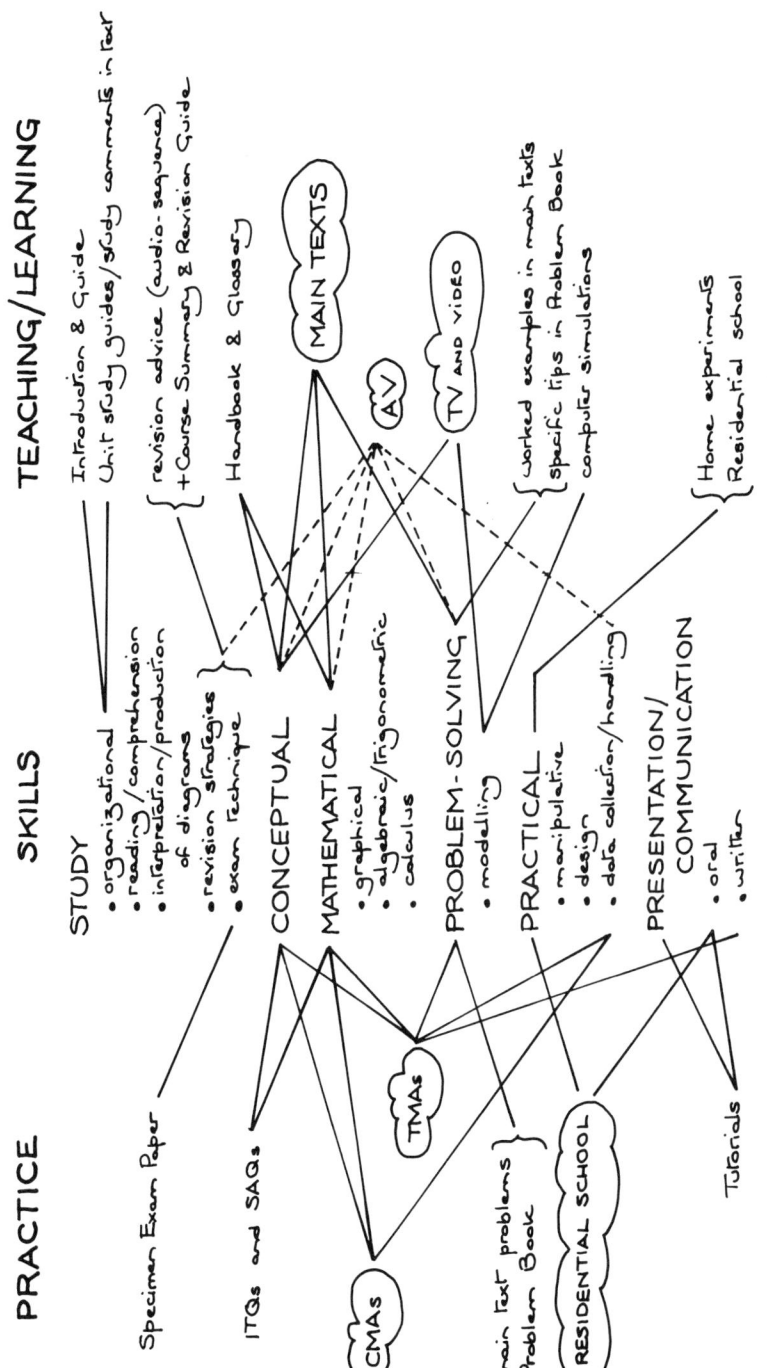

Figure 5.6. Planning diagram for the incorporation of the skills element in the UKOU course *Discovering Physics* (Open University, 1992, *op. cit.*).

5. SKILLS DEVELOPMENT: THREE CASE STUDIES

5.1 Study skills

There is sometimes a perception that study advice is most useful to learners in the initial stages of a programme of study, and that experienced learners have usually evolved 'good' (or at least successful) study strategies. That this is not always the case is illustrated by the following comments from a UKOU student who came to distance learning after already achieving a good degree from a very prestigious 'conventional' university:

> But how did I manage to come through it with so little grasp of what learning actually is, let alone how to go about it? ... Then enter the OU ... I shall never forget my amazement on learning that the purpose of reading is not to remember everything but to engage with the text and allow it to interact with and inform my own understanding. Consequently, I arrived at the [OU] exam period with the facts not consciously learned, [but] facts which actually meant something to me.
> (Owen, 1994, p. 11)

It might even be argued on the basis of this kind of anecdotal evidence that the students best placed to take advantage of study advice are not the 'novice learners', but those whose academic background is already sufficiently secure for them to have time to experiment with alternative strategies. There are also some study skills, such as those involved in literature review, that depend crucially on a considerable knowledge base. Thus study advice is usually appreciated by at least a proportion of the students on a course, whatever its level; this is probably especially true of remote learners and of those coming to study through open entry schemes, whose previous study experience cannot be taken for granted. Figure 5.7, from a third level UKOU chemistry course, illustrates that quite basic study advice can be tailored to benefit even relatively advanced students.

5.2. Problem-solving

The ability to solve problems is a vital skill for scientists, technologists and applied mathematicians in particular, and the importance of the gradual development and eventual transfer of this cluster of skills was highlighted by both Gagné and Bruner:

> The results of using rules in problem solving are not confined to achieving a goal, satisfying as that may be for the thinker. When problem solution is achieved, something is also learned, in the sense that the individual's capability is more or less permanently changed. What emerges from the problem solving is a higher order rule, which becomes part of the individual's repertory. The class of situation, when encountered again

may be responded to with greater facility by means of recall and is no longer looked on as a problem. Problem solving, then, must definitely be considered as a form of learning.

(Gagné, 1970, quoted in Watts, 1991, p. 40)

it is only through the exercise of problem-solving and the effort of discovery that one learns the working heuristic of discovery, the more one has practice of, the more one is able to generalise what one has learned into a style that serves for any kind of task one may encounter . . .

(Bruner, 1961, quoted in Watts, 1991, p. 7)

Problem-solving is, however, a multi-layered skill, and therefore one that is difficult to foster within a distance-learning environment, where group work and face-to-face contact with instructors are at a premium if they occur at at all. One UKOU course in basic physics recently set out to tackle this issue and adopted a problem-solving approach which was introduced from the outset, and pursued throughout the course in worked examples, student self-assessment exercises, assignments and the examination. This approach hinged on giving students a framework, involving clearly identified stages of 'preparation', 'working' and 'checking', on which they could hang their own analyses and solutions of problems. This protocol is described in Figure 5.8.

By focusing on problem solving in this way, the course aimed firstly to assist students in building their understanding, and secondly to consolidate that understanding through the application of basic principles to the analysis

---

4 **Effective reading** Later parts of the Course call for a considerable amount of careful analysis of Source Book chapters, offprints, etc. You may be quite experienced in this respect already, but in case you are not and would like some guidance, we offer the following suggestions for practice in reading the File offprint 'Why bother with a suntan?'

(i) Set down some relevant questions you hope the article will answer. For example:

What happens in the tanning process? Is it harmful in the short term?—the long term? What is its effect on the skin? What molecular constituents are affected? What do they change into, and why are such changes harmful?

(ii) Start with the intention of being able to summarise the main points of the article afterwards (do this for 'Why bother with a suntan?' and then compare your summary with our notes at the end of this section—page 4).

(iii) Note passages that are both difficult *and* relevant. Write down the best understanding you have of their meaning and the reason why they are not completely clear. For example:

Paragraph on effects on DNA

Means(?): Structure of DNA changes so that it cannot perform its function—hence cell dies, or becomes cancerous.

Problems: Forgotten DNA structure—what does thymine dimerisation mean? Why does this accelerate ageing?

(iv) Make a conscious effort to distinguish points of fact from opinions. For example:

Fact: Melanin is produced by exposure to u.v. radiation and it also absorbs u.v. radiation.

Opinion: Melanin absorption of u.v. radiation results in cell damage.

(v) Try to recognise the assumptions behind each important argument. For example:

Argument: Psoralen preparations are dangerous tan-promoting products.

Assumptions: 8–MP and 5–MP act in similar ways. Use of 5–MP could lead to drug-resistant, harmful microorganisms on the skin.

Figure 5.7. Study advice on effective reading from a third level UKOU course *Photochemistry: light, chemical change and life* (Open University, 1991, *op. cit.*, Appendix 3.1, original emphases). Although this course is at honours degree level, the course team still considered it appropriate to include study skills material, covering topics such as model strategies for tackling assignments, analysis of original source material, and essay-writing.

General advice on problem-solving is given in the Units (especially the AV sequences of Units 1 and 2). You should recall that the solution to any problem can be split into the stages of Preparation, Working and Checking. The Preparation stage consists of reading the question, and noting down what is known and what is to be found. If the problem consists of a number of parts, you may want to tackle one part at a time; however, if the parts seem to be closely linked, it's better to get an overview of the whole problem by letting the preparation stage cover all parts of the question. It's often a good idea to draw a diagram at this stage and to introduce symbols for various quantities. You should also try to collect together a series of equations or principles that seem relevant to the problem. Don't worry yet about whether or how you are going to use these equations. The idea is to gather together some resources which, with any luck, will be sufficient to solve the problem.

The dividing line between the preparation and working stages is not precisely defined. It occurs after you have gathered enough tools to tackle the problem but before you get too immersed in algebraic details.

The Working stage constitutes the main part of your solution. It is important to form a plan of attack. Ask yourself how can you link the *unknown* quantities that the question is asking for to the *known* quantities whose values are given. Try to find a chain of equations that make the links. The equations should obviously contain the required unknown quantities and some (usually all) of the known quantities. They may also contain additional unknown variables that will have to be eliminated. You will need at least as many equations as the total number of unknowns. Then you will need to form a plan for eliminating the unwanted unknown variables. With a clear plan, it should not be too difficult to use basic algebra and arithmetic (aided by a calculator) to obtain the required answers.

The final Checking stage is sometimes overlooked but actually very important. Instead of just hoping you have done everything correctly, you should actively check that your answer makes sense. There are several ways of doing this. The simplest way is to check the units of your answer, which may involve some manipulation. For example, is it reasonable for an energy to be a certain number of $kg\,m^2\,s^{-2}$? (The answer is 'yes', but to see why, you will have to remember that kinetic energy is given by the formula $\frac{1}{2}mv^2$.) Another method is to look at a simple limiting case — when a certain quantity is set equal to zero, for instance, does the answer make obvious sense? A slightly different method is to look at the trends in your answer. If a particular quantity is increased, does your answer change in a reasonable way? In some cases, you may be able to use a different physical principle to check your answer. Perhaps you obtained your answer by Newton's laws and can check that it is consistent with the law of conservation of energy.

Figure 5.8. Part of the introductory section of the Problem Book associated with a UKOU second level course *Discovering Physics*, which sets out a problem-solving protocol for students to follow (Open University, 1995, p. 2).

of real-world situations. All the problems were set within a physics context, and none was truly open-ended since all admitted a mathematical solution. Nevertheless, they allowed students to develop skills – such as the evaluation of evidence, the identification of variables, the use of approximations, the construction of appropriate models, and the exploration of different approaches to a given problem – that underlie problem solving in many areas. By featuring these kinds of skills explicitly, it was hoped that students' awareness of good problem-solving techniques would be heightened, and that they would be able to transfer their problem-solving competencies to other science courses or even to other disciplines.

5.3. Communication skills

In programmes with a modular structure, it is not possible for every course to include specific development activities associated with each and every skill that it might be desirable for students to acquire. The best way of building students' skills repertoire under such constraints is to co-ordinate the provision across a suite of courses, with each course emphasizing a different set of skills. Thus when the UKOU produced two new physics courses at the same time, one, as described in Subsection 5.2, concentrated on problem solving, while the other structured many of its activities round the development of communication skills. Like problem-solving, communication is a complex skill, and one that can be difficult to address when learners are remote from both their peer group and their instructors. Figure 5.9 refers to just two of the methods used in this particular course to help students communicate more effectively – an audio-tape highlighting issues, good practice and pitfalls, and a piece of written work on which tutors would give feedback related mainly to the way in which the information was presented, rather than on the information itself.

Spoken communication was similarly covered in a later audio-tape and assignment, with students having to plan a talk, and submit to their tutor an outline, sketches for accompanying visual aids (slides or overhead projector transparencies) and a recording of the first five minutes of the talk. Tutors were instructed to grade and comment on this with particular reference to such issues as: an arresting beginning, clear signposting, use of natural spoken language, avoidance of jargon, imaginative use of visuals, originality, and the liveliness and enthusiasm of the presentation.

A related skill is the analysis of scientific journalism, reports or papers. The increasing trend towards courses that are issues based, rather than concept led, highlights the importance of being able to disentangle scientific facts from sensational or biased presentation. Many interesting exercises for students, at a wide variety of levels, can be built on offprints from the popular press, large circulation magazines (e.g. *New Scientist, Scientific American, Sky and Telescope*) or information bulletins from various agencies; the use of these kinds of material adds immediacy and context to coursework. At the other end of the spectrum, the art of reading papers in the scientific literature often consists in

---

This question relates to Block 1, in particular the skill of writing a non-technical account of a scientific topic. It carries 25% of the marks for this assignment.

The skill of writing a non-technical account of a scientific topic has broad applications. To make it more specific you have to imagine here that you are writing an article for your local newspaper — **please listen to audio band 1, *Writing a newspaper article*, for general advice before you tackle this exercise.** However, it is important for you to realize that the skills that this exercise will help you to develop are relevant to many other sorts of written communication of a scientific topic to a non-scientific audience.

For your article, marks are available as follows:

- *5 marks* for the evidence it contains showing that you understand the topic,

- *20 marks* for its effectiveness as a piece of written communication.

The article should address ONE of the following options. In either case, it *must not exceed 500 words.*

Option 1

On 11 August 1999 a total solar eclipse will be visible from Cornwall, lasting a little over 2 minutes. Imagine that you are writing an article to tempt people to make the journey to see it. You will need to explain what causes such an eclipse, and explain what will be seen — notably solar prominences and the corona. You must warn people not to look at the Sun until the eclipse is total. Mention the British weather!

Figure 5.9. Part of an assignment for a second level UKOU course *Astronomy and Planetary Science*, which specifically addresses the development of communication skills.

being able to distill the essence from a morass of highly technical detail. This kind of skill can only be acquired in tandem with a development in knowledge base, and relates back to the study skills example quoted in Subsection 5.1.

6. GROUP WORK

Another aspect of communication skills is the ability to put one's point across in a group setting, and to work co-operatively with others. There are, of course, particular obstacles to be overcome in trying to develop these kinds of skills in remote learners, but even in the conventional sector many instructors report difficulties in managing group learning. There appear to be a number of possible reasons underlying students' reluctance to work in groups. It may be a specific manifestation of learners' general unwillingness to work in ways that are not in accordance with their expectations. This explanation is borne out by the fact that there does appear to be a cultural element in resistance to group work. The problem may be exacerbated if learners bring to their studies a personal agenda strongly orientated towards subject or qualifications; under such circumstances, it can be difficult to convince them of the relevance of co-operative working skills. We will return to the discussion of these issues in Chapter 13, where we consider the role of the tutor in distance education. Another related problem is student dissatisfaction with group, as opposed to individual, assessment, and ways of negotiating greater acceptance of this will be explored in Chapter 11.

The issues associated with group working are to an extent implicit in all educational programmes designed to foster skills development – the importance firstly of getting students to reflect on and accept not only what they need to *know* but also what they need to be able to *do*, and secondly of giving them a structure in which to work, with clearly defined objectives and assessment criteria.

6

TEXT-BASED LEARNING MATERIALS

Teaching a learner through self-instruction is different from any other kind of teaching . . . even the kind of writing required is quite different from anything most teachers and trainers will have done previously. It is not like writing lecture notes, journal articles, training manuals or textbooks.

(Rowntree, 1990, p. 81)

1. THE ROLE OF TEXTS IN LEARNING AT A DISTANCE

In this chapter we present a description of the considerations that weigh on the designers of distance teaching science texts in facilitating student learning.

In most higher education distance learning systems print is the key teaching medium (Kaye and Rumble, 1981) but different institutions have come to different conclusions about how this text should be produced. The UKOU's model has been to publish teaching texts as typeset booklets or even complete books, but many other distance teaching institutions have chosen different routes. For example, Athabasca University produced loose leaf material in ring binders which could easily be updated, while Deutsches Institut für Fernstudieren (DIF) began by taking the middle path of offering simply produced texts but incorporating good design, filmset headings and professional illustration. All institutions have been changing their production methods with the advent of electronic publishing, but in this chapter we are much less concerned with publishing techniques than with focusing on the *instructional* design of such texts.

The UKOU's text style arose out of early plans which envisaged the production of a small publishing programme of study guides for existing textbooks. However when work started on the production of the first courses, it became clear that the requirements to be met by the texts – taking unqualified students to degree level – could not be fulfilled by existing textbooks. So it was decided that teaching texts should be specially prepared – and that they should be of good quality in terms of both the teaching material and its presentation. Waller and Lefrere (1981, p. 36) write of the need for the distance teaching institutions to establish the quality of the material they produce and thereby distinguish themselves from 'the image of mail order diplomas, crammer courses, tatty duplicated lecture notes, advertisements offering increased earning power and testimonial letters'.

The importance of public image in the decision to use real printing rather than office reprographics led to the emergence of the Open University as a major educational publisher. Indeed, the University's publishing division operates just like a conventional publisher but replaces the profit motive with that of maximizing educational quality.

Kaye and Rumble (1981) have produced a good summary of the characteristics of the medium of print – which is reproduced in Figure 6.1. They point out that many different functions have to be served by the print, not just to present content, impart facts or develop skills, but to supply links out to other media, and to provide information on how the course is to be studied. It is clear from Figure 6.1 that as a medium for the delivery of educational material to remote students, print has enormous logistical advantages. Texts are comparatively easy and cheap to design, reproduce and deliver. For students in domestic or workplace environments, or for those on the move, print is usually the most convenient medium to work with, being self-contained, portable and easy to access.

Arguably it is the logistical advantages of print, rather than purely educational considerations, that have made text the main instructional medium in most distance teaching systems. But as Laurillard has pointed out:

Logistics change with technological and cultural changes . . . so we have to be clear about the true extent of the pedagogical characteristics of print to be able to judge these against its changing comparative logistics.
(Laurillard, 1993, p. 109)

Some of the characteristics of the text-based medium have long been recognized and valued by instructors. There is a considerable element of student control: texts can be skimmed, worked through carefully at the student's own pace or dipped into for reference or revision. In these respects, print may be regarded as a more student-centred instructional medium than mass audience media such as lectures or television programmes. However, print also has drawbacks, especially if it is the sole or primary means of providing an educational experience. Chief among these is the fact that print can never be interactive; it is not a medium in which students can receive individual feedback on their actions. There are difficulties for the instructor as well: he or she does not witness the students' reactions to the material, and cannot easily alter it to take up issues or problems that the students may raise unexpectedly. A major concern for instructional designers in distance education is to find devices that will minimize as far as possible the deficiencies of text-based materials and provide a quality learning experience for remote students.

Arguably, a lecture is closer to distance teaching style than is a textbook, partly because it is usually less formal. But the ideal 'model' for the distance teaching text is the good tutorial, where the tutor involves the students in the argument under construction by making opportunities for student participation. The author can check the students' understanding of what has been already said and what they think about it, asking students to anticipate where the argument is leading, and getting them to contribute their own experiences. A student in such a tutorial needs feedback about the answers he or she gives. This effect can be achieved in print by a judicious

Types of material	Specially written correspondence texts or lessons Text books and readers: specially written or already published Supplementary items: notes on broadcasts, assignments, instructions, drawings, photographs, maps, charts Journals, newspapers, periodicals Reading guides, bibliographies
Pedagogical functions	To impart facts To develop skills To illustrate how knowledge can be organised for learning To provide links to tutors/other students
Motivational functions	Student can work at own pace Self-assessment questions can provide reinforcement Provides a permanent record for revision Academically respectable Written comments from tutors provide reinforcement
Demands on students	Fairly high levels of literacy required Motivation and/or previous experience of independent learning from reading needed Tends to impose a linear learning strategy
Flexibility and costs	Generally the least expensive and most flexible of the media Must be prepared well in advance for use by students Major updating and revision can be costly, but brief supplementary items (notes, errata, etc.) can be prepared quickly in response to student/tutor feedback
Creation, production and distribution	Special skills required for preparation of self-instructional written material (implications for staff training, use of consultants) Technical production skills needed: editing, design, illustration, printing, storage Distribution arrangements; post, bulk deliveries to local centres, road/rail, air transport

Figure 6.1. Characteristics of print media, as used in distance education (Kaye and Rumble, 1981, p. 51).

use of exercises such as self-assessment questions. We will return to ways of designing 'student active' texts in Section 3.

Rowntree has coined the term 'tutorial in print' to capture the idea that the material produced in print for self instruction has to carry out all the functions a teacher would fulfill in a conventional situation:

> guiding, motivating, intriguing, expounding, explaining, provoking, reminding, asking questions, discussing alternative answers, appraising each learner's progress, giving appropriate remedial or enrichment help ... and so on.
>
> (Rowntree, 1990, p. 11)

To help those writing distance teaching texts, he goes on to suggest that the author imagines the learner not as part of a lecture audience or as a textbook reader whose teacher is at hand but as an individual that he or she is tutoring for a few hours. Everything the remote instructor might want to say to this individual needs to be written down to form a tutorial in print. So text is used not just to carry information about the subject under study but also to simulate tutorials and to organize other aspects of the course such as experiments or projects.

Print is of value not only to distance teachers but also to conventional teachers who rely on students' use of textbooks. However it is important to remember that a distance teaching text is not the same as a textbook. One way of thinking about the difference between them is to consider what Norbert Sanders says about textbooks:

> Although many are attractive, accurate, readable and understandable, they are one of the biggest deterrents to thinking ... The key function of the writer is to explain and a good explanation is interesting, orderly, accurate and complete ... The textbook is weak in that it offers little opportunity for any mental activity other than remembering.
>
> (Sanders, 1966, p. 158)

Distance teaching texts do look different from conventional textbooks. Sometimes it appears that the physical appearance of a self-instructional text is simply a matter of taste or at best a question of aesthetics, and attention to detail of presentation referred to disparagingly as 'cosmetic'. Some research has been done into factors affecting the educational effectiveness of teaching text, but much of it tends to focus on specific issues such as legibility of typefaces rather than on 'higher order organizational problems' of textual material. One example of research based on presentation issues is that undertaken by Hartley and Burnhill (1977a, p. 69), who summarize evidence on the best page size and column structure ('single column on A4 better for text broken by tables and figures'), emphasizing devices ('bold lower case or underlining rather than capitals') and type size ('10 pt on 12 pt line to line feed, 8 pt on 10 minimum').

A key question is whether the students can understand the words used in a written text. Research suggests that readability measures are notoriously unreliable predictors of the comprehensibility of a text (see e.g. Klare, 1981;

Duffy, 1985). This is because comprehension depends on the interaction of the reader with the text. It therefore brings into play graphic interpretational skills and the prior knowledge of the reader, as well as reading skills. A meaningful readability score for a text thus requires a test of the reader as well as an account of text features. However, readability scores are in wide use. For example, in the USA many military training manuals are designed on contract with specified targets for readability. Rowntree (1990) recommends using more rough and ready tools, accepting their imperfections but believing that they help and improve an author's judgement of his or her own prose. He describes his own Complexity Quotient, and also Gunning's Fog Index (1952). These build on the idea that readability depends on a friendly conversational tone, human interest, well chosen and preferably short words, short sentences and specific concrete examples.

One of the main problems for the remote student can be maintaining their interest in the material when deprived of partners in studying. In other chapters we will describe other ways of overcoming this problem, but in relation to print material involving the student in active study of the material is one way of helping. The problem of motivation extends even to the distance teacher. Riley (1984a, b, c) conducted a study of the problem of drafting distance teaching materials by interviewing some authors who were members of a number of UKOU course teams over the period of course preparation. Among the factors that influenced how drafts were produced she identified the important role played by such private factors as the individual author's fears and anxieties. A scientist wrote to her of his difficulty in settling down to write as follows:

> It leads me to procrastinate . . . to displacement activities. It's OK once I'm doing it, I can write very fast when I'm happy about what I'm doing and very slowly when I'm not. It may actually influence what's finally in the unit, the quality for the students must be affected by my reluctance and lack of confidence.
>
> (Riley, 1984a, p. 235)

Riley's conclusions were that authors' personal attitudes had a significant influence on course design, and that the 'private' processes of drafting and criticism needed as much attention as the more 'public' concerns of subject matter and instructional design. We will return to this point in more detail in Chapter 12.

2. DESIGNING INSTRUCTIONAL TEXTS IN SCIENCE

As well as the general drawbacks to print-based educational media, there are some particular problems associated with teaching or learning science through texts. One difficulty, that of the language of science, has already been alluded to in Chapter 2. 'Ordinary language, technical terms, diagrams, equations and special notation, this is the apparatus used by scientists to describe subject matter' (Macdonald-Ross, 1977, p. 69).

Another problem is that clear writing about science depends on the construction of a logical argument, yet the mere fact of seeing such an argument in print can deaden the notion that any debate about the topic is possible. The linear structure of text-based instruction and its non-interactive nature can compound this problem, leading students to equate 'mastering the subject' with 'learning a set of facts and sometimes skills', rather than seeing this as a process to which they contribute their own ideas.

As noted in Chapter 2, science is a hierarchical subject, and therefore the careful structuring of the text material can help learners. For example, specification of expected knowledge and skills on entry, and the provision of graphical aids such as concept maps or networks to highlight the internal structure of the teaching material are useful devices and will be considered in more detail in Section 3. However flexibility, which can be one of the chief virtues of self instruction, can disappear as a result.

A concept map is only one of the many ways that graphics can be introduced into text. Diagrams, graphs, drawings, paintings, photographs and maps can all be useful and will be considered in more detail in Section 4. Crooks, Rowntree and Waller (1979) classify the use of such devices as either showing what things look like, showing how things work, making an impact or displaying quantitative data. Research on how to present quantitative data has been distilled by Macdonald-Ross into the following advice for graphic designers (1977, pp. 68–9):

- for showing general relationships to the general public (or first or second year tertiary students), use horizontal charts;
- for showing general relationships to the professional (or advanced tertiary students), use tables;
- for showing exact data on experiments or investigations, use graphs or tables;
- for showing data for operational purposes, use tables.

This list illustrates the distinct nature of writing about science. Students will find that they have to relate sentences describing phenomena to information presented in diagrams, tables or graphs, and to deal with a mix of quantitive and qualitative information. In any science text, but particularly in those designed for independent study, it is therefore helpful to learners to provide devices that bring out the structure of the text, and encourage active engagement with the material. Such devices, which may be grouped together under the term 'access structures' are the subject of the next section.

3. ACCESS STRUCTURES

One way of defining access structures relies on the fact that they are usually highlighted simply by the layout of the material. Such devices are therefore seen to operate on the basis of 'coordinated use of typographically signalled structural cues [that] can help students read texts using selective sampling strategies' (Waller, 1979, p. 175).

Some of the most obvious types of devices covered by the definition include title pages, contents lists, headings, indexes and glossaries. The value to learners of these kinds of structures has been surprisingly little researched, perhaps because they are seen as being so clearly useful. Hartley and Burnhill (1977b) have also suggested that what early research there was may have been flawed because it was based on uncritical acceptance of existing typographical conventions, did not use materials in context and had no theoretical base. A more pragmatic approach was also advocated by Macdonald-Ross and Waller (1975), who considered that research into the organizational value of typographically signalled structures should be focused on the testing of alternative strategies used in genuine learning situations. Some new versions of long-established structures have also been evaluated, and the comparative nature of such studies gives them similar validity (see e.g. Zimmer, 1984).

For the author of instructional material, a more useful way of classifying access devices highlights not their typographical characteristics but their role in the teaching and learning process. Four types of structure are of particular interest:

- preliminary organizers;
- objectives;
- activities or inserted questions;
- retrospective organizers.

Preliminary organizers

Preliminary organizers are devices designed to introduce learners to a piece of instruction. One specific type, the so-called advance organizer, was as mentioned in Chapter 3, promoted by Ausubel who described its characteristics as follows:

> These organisers are introduced in advance of the learning material itself, and are also presented at a higher level of abstraction, generality, and inclusiveness than the learning task itself: . . . this strategy . . . satisfies . . . the criteria for enhancing the organisational strength of cognitive structure. Summaries and overviews on the other hand, are ordinarily presented at the same level of abstraction, generality and inclusiveness as the material itself. They simply emphasize the salient points of the material by omitting less important information, and largely achieve their effect by repetition.
>
> (Ausubel, 1963, p. 81)

Ausubel himself produced much of the empirical evidence for the effect of advance organizers on student learning. Mayer (1979) has shown that they are particularly useful for learners in mathematics and science.

However, not all types of preliminary organizer satisfy Ausubel's rather restrictive definition of an advance organizer. Other kinds, identified by Nathenson and Henderson (1980) and Melton (1984) include:

- lists of terms, which distinguish between those assumed to be already understood by the learner and those whose meaning will be developed in the text.
- concept maps, whose function is to bring out for students the relationships between the concepts that are to be covered in the text (or other associated material). A typical concept map is illustrated in Figure 6.2.
- study guides, which explain to students how the learning materials are structured, and give some suggestions for how they could be approached. Figure 6.3 shows one way of constructing a study guide.
- motivational introductions, which aim to demonstrate to students why the material is relevant and worth studying. An example of such an introduction is illustrated in Figure 6.4.

Objectives were also considered by Nathenson and Henderson to be a type of preliminary organizer. However, as will be shown, there are good reasons for considering objectives as a separate class of access structure.

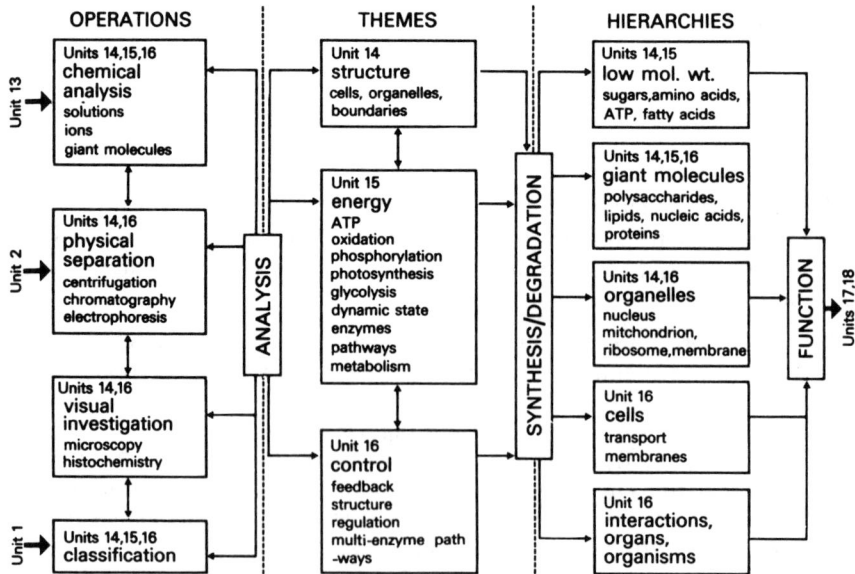

Figure 6.2. Conceptual map for part of the UKOU Science Foundation course, dealing with the interfaces between chemistry and biology. Links between 'units' are made on several different levels, so guiding students to make connections in a variety of different ways (Open University, *Science Foundation Course*, S100).

STUDY GUIDE

After the introduction (Section 1), this Unit is divided into four main sections. Sections 2 and 3 develop the important concepts and equations used to describe magnetic fields. Sections 4 and 5 are mostly concerned with the applications of these principles to a variety of situations, and the TV programme illustrates the principles with particular reference to non-uniform fields. To derive maximum benefit from the programme, you should watch it towards the end of your work on this Unit. In any event, you should read Sections 2 and 5.2 before watching the programme.

Table 1 gives some advice on how to apportion your time to the different sections.

Throughout this Unit, you will be required to visualize three-dimensional fields, although these can only be given two-dimensional representation on the flat pages of the text. You may find it helpful to arm yourself with some odd lengths of wire that can be twisted into loops or coils, as an aid to interpreting the diagrams.

Table 1
A suggested division of study time

Section 1	Read quickly
Section 2	15%
Section 3	15%
Section 4	40%
Section 5	30%

Figure 6.3. Example of a study guide for a unit on magnetic fields. This particular guide refers to all the course components associated with the unit, and explains how they fit together. It also gives learners an indication of how they might sensibly apportion their study time (Open University, *Discovering Physics*, Unit 10, p. 2).

INTRODUCTION

When Albert Einstein proposed the special theory of relativity in 1905, his deductions were based on two postulates — two fundamental principles that he assumed to be true. These two assumptions led Einstein to a new view of space and time, a view that was rapidly accepted by the scientific world. Naturally, a new attitude towards such basic matters as the nature of time and space resulted in changes in many already well-established areas of physics. There was an Einsteinian revolution, the magnitude of which has rarely been equalled in the history of physics.

Although this Unit does not follow Einstein's work in detail, it does have the same overall approach. Apart from the introduction and conclusion, the Main Text is divided into six parts. Sections 2 and 3 explore the first postulate of relativity, while Section 4 examines the second postulate, relating to the constancy of the speed of light. In Sections 5 and 6, these two principles are brought together, and their implications for time and space are investigated. Then, in Section 7, I examine the effect of Einstein's theory on some fundamental aspects of mechanics. As you will discover, the predictions of special relativity become equivalent to those of Newtonian ideas at ordinary speeds; but they lead to very surprising and unexpected predictions at speeds comparable with the speed of light.

Special relativity has the reputation of being difficult to grasp, and yet it is fair to say that, alone among the topics of twentieth century physics, it can be explained honestly and simply using mathematics that involves only basic algebra and geometry. The key to a straightforward (but quite rigorous) development of the basic ideas of special relativity is to adopt a careful and thoroughly logical approach to the experimental evidence and the implications that derive from it.

By proceeding methodically, you may come to appreciate that, for many physicists, the fascination of Einstein's theory comes not only from its extraordinary predictions, but also from the irresistible, unrelenting force of the arguments that lead to these predictions.

Figure 6.4. Motivational introduction for a unit on special relativity (Open University, *Discovering Physics*, Unit 12, p. 3).

Objectives

As noted in Chapter 3, Bloom and Tyler were influential in developing the idea of using objectives to help learners. Sometimes the words 'aims', or 'goals' are used loosely to describe the overall intention or the purpose of a piece of teaching. The term 'educational objective' is used to state more precisely what the outcome of the teaching is supposed to be. Mager (1962) identified the characteristics of objectives and explained in detail how they might be produced in practice, recommending that a statement of a behavioural objective should contain, as well as a statement of the behaviour that the teaching material is designed to realize, both a description of the conditions under which the desired behaviour ought to be visible and an indication of the level of performance that would indicate the objective had been achieved.

Examples of behavioural objectives as used in a science distance education text were illustrated in Figure 5.3. However, it is worth noting that those examples, like the majority of objectives formed for undergraduate learners, do not conform fully to Mager's specifications, in that they give no performance indicators (such as expected standards of speed or accuracy in solving mathematical problems).

The reason that authors such as Nathenson and Henderson (1980) considered objectives to be a form of preliminary organizer was that in the 1970s there was a tendency to put a list of objectives at the beginning of instructional texts. Indeed as illustrated in Figure 6.5, early versions of UKOU courses started with rather vaguely stated aims (essentially of what the *author* intended to do) immediately followed by more detailed objectives (listing what the *student* should be able to do after working through the material).

However research into the effectiveness of such objectives (see e.g. Melton, 1978) suggested that incidental learning might be depressed by their use, and opened a debate about whether precisely stated objectives placed before the instructional materials could be meaningful to learners. The current view, exemplified in some UKOU science texts, is that objectives, as illustrated in Figure 5.3, are best located at the end of texts, and linked to a series of sample self-assessment exercises that allow students to measure their own achievement. The provision of self-assessment items also ties in with the view of some instructional designers that behavioural objectives should include a specification of the way in which the desired behaviour will be measured.

Activities

Students also need to know when they are working through an instructional text, whether they are understanding properly what they read. Self-assessment questions, like those in Figure 6.6, can be given at the end of each section of text, with comments and model answers. Their function is to help students

Aims

In Part A we intend to:

1 Show how the standard cosmological model can be supported and strengthened by interpreting the thermal radiation which has been detected at radio (microwave) frequencies as being a relict of the radiation generated by the Big Bang.

2 Show how the existence of this radiation opens up the possibility of measuring the present velocity of the Earth with respect to the rest frame of the Universe at the time the radiation was formed, and may in the future indicate answers to the problems of galaxy formation, and of the conditions in the Universe during the first seconds of its existence.

Objectives

After reading Part A you should be able to:

1 Give an account of the likely stages in the production of the 3 K radiation.

2 Describe how thermal radiation is affected by the expansion of the Universe.

3 Explain how to specify motion in the Universe and how observations of thermal radiation are affected by the motion of the observer with respect to it.

4 Give some alternatives to the standard cosmological model and say to what extent the existence of the 3 K radiation limits their credibility.

Figure 6.5. Aims and objectives for a portion of an early UKOU third level course on general relativity and cosmology. The differences with the formats adopted in Figure 5.3 are quite striking (Open University, *Understanding Space and Time*, S354, Block 6, p. 6).

---

concentrate on the main points in the text and to provide them with instant feedback on their mastery of the material.

The combination of objectives, activities and self-assessment questions changes the way that students perceive the text, encouraging them to develop their own conceptions of the material by engaging in purposeful activity. It is not possible for there to be an enormous amount of feedback provided to students on the activities, although limited text can be devoted to deal with predictable wrong responses to in-text questions. In a distance education system, most of the feedback has to come in other ways, from more interactive media such as computer tutorials and from the course tutors either in face-to-face sessions or through comments on assignments. One way for the author of the text to deal with this lack of feedback is to research the possible prior conceptions that students may have and incorporate ways of addressing these in the instructional strategy.

Evidence on the effectiveness of self-assessment questions in text is somewhat equivocal. Lockwood (1992) reports that in general only a small proportion of students actually write anything down when asked to do so for an activity. Scanlon (1984), however, found that 98 per cent of UKOU students made use of the objectives in the science foundation course teaching texts, while 87 per cent found them useful; 99 per cent of students found the self-assessment questions very or fairly useful. Waller (1984) investigated a structured exercise on the Galapagos finches from the UKOU science foundation course, a sequence of in-text questions which guided students to develop a hypothesis about evolutionary processes. He demonstrated that students learned more from this than from an equivalent narrative version with no in-text questions. Waller also found that while the majority of students involved in this study felt that such exercises were useful, they disliked 'short cryptic answers which failed to predict and diagnose the source of incorrect answers' (p. 164).

A number of attempts have been made to develop further the use of activities in text so as to increase the amount and success of student interaction with the material. One worry with the use of badly designed or trivial activities in text is that they might encourage 'surface' processing of texts as opposed to a 'deep' or holistic approach among students (see e.g. Marton and Säljö, 1976a, and b). We referred in Chapter 3 to the importance of the students' orientation to study and the importance of their perception of the processes of teaching and learning. It is vital that activities be designed in such a way as to encourage students to consider the structure of the argument in the text and actively to use concepts that they are being taught, rather than just skimming the material for key words and phrases, or substituting in numbers. Of course,

---

The following SAQs will demonstrate how the law of conservation of energy is applied to the human body.

SAQ 17   If the average basal metabolic rate of an individual is 7.0×10^6 J per day, approximately how many people at rest in a room would be necessary to emit heat at a rate equivalent to a 1 kW electric fire?

SAQ 18   While on my latest diet, I secretly ate an extra packet of plain crisps. The label on the packet stated that the energy content was 567 kJ. When my wife found out, she told me I must use up this energy by going for a run. If, when running on level ground, my average power consumption is 800 W and my speed is a constant $4.0 \, \mathrm{m\,s^{-1}}$, estimate the distance I have to run to ensure that I have used up all the energy contained in the crisps. Why might I get away with running a shorter distance? When I stop running, my kinetic energy is reduced to zero. Where has this energy gone?

Figure 6.6. Self-assessment questions (SAQs), designed to allow students to check their understanding of the concepts of power, energy and energy conservation (Open University, *Discovering Physics*, Unit 3, p. 37).

activities are not confined to promoting *conceptual* development in students. Most activities are strongly process oriented, and are a vital element in the acquisition of skills.

A slightly different use of activities in text is the diagnostic pre-test. This can be self-administered by students, and may be used to direct them into preparatory or supplementary material as appropriate for their individual needs. Typical examples of remedial material for science students that might be packaged as supplementary texts include basic mathematics, statistical analysis, and the rules of nomenclature in organic chemistry. Ideally, such material should itself also incorporate activities! In this way, the adaptivity of the main teaching text can be enhanced, allowing for a wider range of starting points among the students.

Retrospective organizers

The most common type of retrospective organizer is the final summary, in which the author usually tries to provide an overview for the learners by condensing the most important points into a short précis. While summaries are most often used to encapsulate the main messages of quite large pieces of instructional material (at the level of a chapter, say), more frequent (e.g. end of section) summaries can be of particular value to students, especially if accompanied by self-assessment post-tests which allow learners to check their understanding before moving on to the next section.

The approaches to access devices that are common in distance education are almost totally instructor controlled: it is the instructor who sets the agenda, assembles the resources, structures the text, directs the way in which students are expected to work with the material, designs the activities and supplies the answers by which students assess their progress. There is a sense in which this degree of control is reassuring for the instructor, who may perceive it in some ways as a compensation for direct contact with the learners. However, the feel of control may in large measure be an illusion. As Marland and Store (1982) have pointed out, students read texts selectively, and can exercise a veto over the use of assess structures they do not find helpful. Good instructional design should recognize this power of veto.

4. THE ROLE OF ILLUSTRATIONS IN SCIENCE TEACHING TEXTS

There is a growing body of research evidence suggesting that illustration is a very important part of print-based learning materials (see e.g. Mayer and Gallini, 1990 and references therein). Various classification schemes for illustrations have been proposed, but a simple clustering by function is:

- decoration/visualization: typical examples would include cartoons to enliven otherwise solid or difficult portions of text, portraits of famous scientists, pictures of historic experiments, landscape photographs showing a lava flow or a glacial moraine (to accompany a geology text), etc.

- organization/transformation: diagrams to help the learner arrange information into a meaningful structure, or to remember key material. Concept maps, as illustrated in Figure 6.2, are obvious examples of illustrations within this cluster. Another familiar example is the periodic table of the elements.

- interpretation/explanation: illustrations that promote understanding of the text. Some examples are shown in Figure 6.7.

FIGURE 21   (a) A typical small-scale normal fault in layered sedimentary rock. The fault is in the centre of the photograph and runs from top left to bottom right. The rocks on the right hand side of the fault have moved down relative to those on the left of the fault, by a distance of about the length of the geological hammer, say 30–50 cm. (b) This is a block model showing the formation of a normal fault. Compare this with the fault in (a).

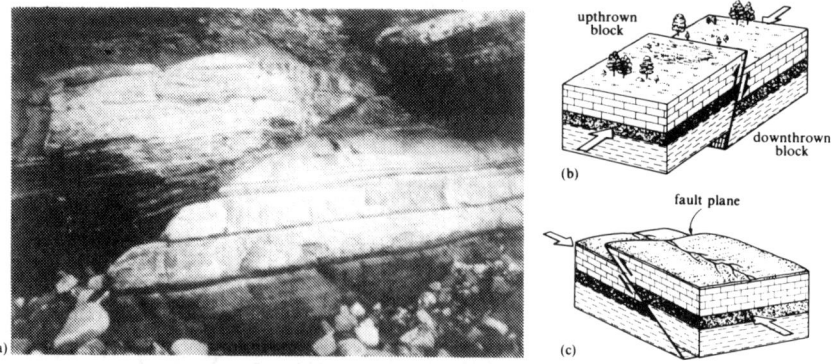

FIGURE 22   (a) A typical small-scale reverse fault, which you should compare with Figure 21a. The rocks on the left-hand side of the fault have moved *up* over those on the right of the fault.
(b) Block model showing the formation of a reverse fault. Compare this with the fault shown in (a).
(c) Block model of a thrust fault, where the angle that the fault plane makes with the horizontal is much lower than for the reverse fault in (a) and (b)

Figure 6.7. Examples of explanative illustrations, showing geological structures resulting from tension and from pressure (Open University, *A Science Foundation Course*, Unit 27, p. 48).

This third cluster has been investigated by Mayer and Gallini (1990) in a study designed to find out how illustrations can be constructed and integrated with text in order to promote learners' understanding of the working of scientific devices (the examples they selected were the operation of a braking system, a pump and a generator). They tested the effectiveness of three types of diagram: 'parts' illustrations that illustrated the system topology and labelled the various components, static 'steps' illustrations that included labels for each major action, and dynamic 'parts-and-steps' illustrations that showed alternative states of the device with both parts and actions labelled. The conclusions from this study were summarized as follows:

> Results indicated that parts and steps (but not the other) illustrations consistently improved performance on recall of conceptual (but not nonconceptual) information and creative problem solving (but not verbal retention), and these results were obtained mainly for low prior knowledge students.
>
> (Mayer and Gallini, 1990, p. 175)

This and other similar studies in cognitive psychology, emphasize the potential and importance of integrating visual instructional material into textual media in order to facilitate learning of science concepts and processes.

5. PRODUCING DISTANCE LEARNING PRINT MATERIAL BY CUSTOMIZING EXISTING MATERIAL

One solution to the problems of constructing open learning materials or distance teaching texts from scratch is to make use of a previously published textbook and to produce 'wrap-around' materials that turn it into a distance teaching text. In the UKOU this procedure is only followed in higher level courses, where students can be expected to have developed the necessary study skills to cope with the rather syncopated form of presentation. This approach was adopted in a third level UKOU electromagnetism course, where a well-respected student textbook was incorporated as a major element and study guides written to lead students through the course materials, which included television and video presentations. Figure 6.8 is an extract from one of the study guides which functions as a wrap-around workbook. This illustrates that the main purpose of a guide for self study of a 'conventional' textbook is to tell learners *how* best to use the text in their own context. One key task is to divide the material to be studied into manageable portions. The wrap-around material also might include some or all of the following: information on objectives for students to achieve through their study, introductions, section study guides and advance organizers, summaries, additional material like alternative exercises and examples, activities (with associated feedback), indications of sections of the textbook which are not relevant, and additional or alternative explanation

---

Now read GP from the beginning of 1 to the end of 1.1.

Summary Electric charge occurs in two types, positive and negative, and its unit of measurement is the coulomb. Like charges repel and unlike charges attract. The charge on the electron, $-e$, is -1.602×10^{-19} C. The inverse square law of force between charges is encapsulated in Coulomb's law (GP Eq 1.2). The total electrostatic force on a charge due to several other charges is the vector sum of all the individual forces.

Comments

1 GP mention the gaussian system of units. In this course, you will be expected to use only SI units.

2 It is important to be very precise with vector notation and the use of subscripts right from the start. In particular, you should be clear about the meaning of subscripts that appear on various quantities (GP Eqs 1.1, 1.2, etc.). The vector $\mathbf{F}_{21}$ means the force *on* particle 1 *due* to particle 2, and the position vector $\mathbf{r}_{21}$ means the position vector *of* particle 1 *from* particle 2, i.e. $\mathbf{r}_{21} = \mathbf{r}_1 - \mathbf{r}_2$. Errors in the use of these subscripts may lead to errors in signs.

3 You will probably be used to writing Coulomb's law with r^2 in the denominator and the unit vector, $\hat{\mathbf{r}}$, defined as $\mathbf{r}/r$. This is entirely equivalent to the convention adopted by GP (Eqs 1.1, 1.2, etc.) of gathering all the r's into r^3 in the denominator. Coulomb's law is still, of course, an inverse square law of force.

Figure 6.8. Extract from study guide for part of a course designed round a conventional textbook. GP are the initials of the textbook authors (Open University, *Electromagnetism*, SMT 356, Unit 2, p. 4).

and illustrations. Wrap-around material can also contain instructions for use of any other media included in the course or for additional practical activities.

Another similar use of text at the UKOU is the production of 'course readers' – volumes of readings to accompany distance learning courses. These are not teaching texts but collections of documents, often previously published research papers or specially commissioned articles from subject experts, which authors of courses can incorporate into the study programme. One problem of distance learning is the inadequacy and uneven distribution of library facilities through which students may gain access to printed materials not supplied as part of their course. When such courses require students to develop skills such as interpreting and drawing conclusions from research articles, i.e. skills involving the use of original sources, course readers can fill the gap.

An alternative approach is to plan from the outset to produce the text component of courses jointly with a commercial publisher so that individual volumes can be sold separately to the general public. One of the benefits of the co-publishing arrangement is that resources can be found to pay for a higher number of illustrations and colour photographs than is often possible in courses produced completely in house by a distance teaching institution. One of the disadvantages of this approach to producing courses in this way is that authors may feel that, as the material will be so widely available to the general public, including their academic peers, it is important to include material, for some notion of completeness, which is not strictly necessary for a student audience. There is evidence that some courses produced in this way can significantly overload students.

Another disadvantage of the co-publishing approach is that, because the books are on general sale, it is not possible to refer out to other components of the course material from within the text. This may make it more difficult for the instructional designers to achieve good integration between the various media used in the course. Most publishers accustomed to producing conventional textbooks will however normally agree to the incorporation of familiar access devices such as in-text, self-assessment questions, summaries and lists of objectives. Because of the difficulties of providing full integration co-published texts are perhaps more suited to higher level courses, on the basis that students on such courses are better able to make links for themselves between different course books and between the texts and other components. From the publisher's point of view, the subject matter for the texts should be one of interest to as large an audience as possible, apart from the registered students on the course. Examples of recent successful courses at the UKOU produced by co-publishing arrangements include a third level course in oceanography and a second level, cross-faculty course on environmental matters.

6. FUTURE DEVELOPMENTS

Despite recent advances in the electronic creation and dissemination of information, print remains a key medium in education, particularly in the distance learning sector. Yet computers have already revolutionized the production of texts. Electronic means of production have made it easier to produce revised versions of print and blurred the traditional roles of author and editor. So, why is there still such a reliance on print media? One reason for this is that for many purposes, print is still the most appropriate vehicle.

Figure 6.9 summarizes the advantages and disadvantages of print. The portability of books together with their integrity, lack of reliance on hardware, and aesthetic appeal is difficult to beat for the distance learner.

However, interest in developing electronic documents has been sparked by the availability of technology such as CD-ROM, which can allow the storage and retrieval of vast amounts of text. Marchioni and Schneiderman illustrate the convenience of CD-ROM as follows:

> The print version of the [Grolier's electronic] encyclopedia occupies 20 volumes. The hypertext version consists of 60 megabytes of text and 50 megabytes of indexes that contain pointers to each occurrence of every word in the encyclopedia, all occupying less than one-fifth of a single CD-ROM disc.
>
> (Marchioni and Schneiderman, 1988, p. 254)

Characteristic	Advantages	Disadvantages
Integrity of information	Historical value Never inaccessible because of unreliable hardware	Readers can never alter content Readers cannot customize information Cannot conform to user preferences (type size, margin width)
Physical entities	Portable Allows browsing and exploring Allows annotation and underlining Aesthetically appealing	Limited to 2-D information Limited to static text and graphics Costly to reproduce for quickly outdated information Often hard to locate specific information
Static		Cannot handle sound or motion Difficult to create multiple indices
Advanced technology	Well-defined and accepted standards Typography. graphic design. and photo reproduction refined fields High-resolution print and graphics Easy to read	Joint authorship difficult Rekeying text is error-prone

Figure 6.9. Advantages and disadvantages of print media (taken from Yankelovich *et al.*, 1985, as reprinted in Boyd-Barrett and Scanlon, 1990, p. 236). © 1985 IEEE.

This has significant implications for the supply of print material to distance learning students. The provision of an original library of articles on CD-ROM is an attractive alternative to the production of course readers.

However, the development of electronic documents for education has been spurred on not just by the availability of the technology to do so, but also by the potential of the medium for allowing both instructors and students to explore connections. This advantage is highlighted in the following definition of hypertext as 'a computer based software system for organising and storing information to be accessed nonsequentially and constructed collaboratively by authors and users' (Jonassen, 1991, p. 83). Documents can include text, pictures, diagrams or even dynamic images such as computer animations and computer-controlled video sequences. Links within documents and between documents allow readers and learners to deal with documents in a way quite different from print. It is obvious that for, say, a student of chemistry the ability to look at a video clip in which a complicated 3D molecule is rotated is likely to be more helpful than a text-based description or a static 2D representation. The advantages of such multimedia documents will be discussed further in

Chapter 9, but much has still to be learned about how readers of hypertext documents use the features available to them, in what way readers learn from such documents and how such systems are best designed.

7. CONCLUSIONS

It is important to realize that however hard authors and designers work it is impossible to produce a text that is universally effective – in the sense of being immediately comprehensible to a wide range of students. It is essential to have a clear idea of the target audience for the text being produced, together with information about design features which address some of the difficulties that distance learners have in learning from text. The hierarchical nature of the subject matter in science means that a carefully sequenced piece of teaching is often necessary. There remains the problem of how to sequence instruction carefully while still allowing individual learners to retain some measure of control over their preferred learning strategies. However the flexibility of new electronic media offers a radically different approach to learning, which we will consider in more detail in succeeding chapters.

7

AUDIOVISUAL MEDIA

What we envisage is the creation of a new educational trust representative of the universities and other educational organisations, associations of teachers, the broadcasting authorities, publishers, public and private bodies, producers capable of producing television and other educational material. Broadcasting time could be found either by the allocation for the fourth television channel together with appropriate radio facilities or by pre-empting time from the three existing channels.

(Wilson, 1963, quoted in Perry, 1976, p. 9)

1. INTRODUCTION

The quotation at the head of this chapter is one of the first public pronouncements by the then British Prime Minister, Harold Wilson, on an idea for what he called 'The University of the Air', in which television broadcasts would play a crucial role. These early plans gradually evolved and consolidated into the UKOU, and although the institution as it exists today is far removed in many ways from Wilson's original conception, the use of broadcast material and other audiovisual media has remained a cornerstone of its instructional strategies. In this chapter we will concentrate on the use of a wide range of audiovisual media in a variety of educational settings but we will also present a brief account of the particular history of the development of audiovisual media at the UKOU.

We will describe the characteristics of a variety of media including television (broadcast or on video-cassette), audio (either broadcast radio or on audio-cassette), audiovision (audio accompanied by some visual material), and the print (text and graphics) which may accompany these media. All video images (whether moving picture, film, broadcast TV, videotape or videodisc) share some common advantages, but in this chapter we will deal simply with broadcast TV and videotape while videodisc and other more recent forms of multimedia are discussed in Chapter 9.

2. AUDIOVISUAL MEDIA AS TOOLS FOR EDUCATION

Broadcast TV

Broadcast TV has played a large role in the development of distance education. The quotation which heads this chapter is a reminder of its importance in the planning stage of the UKOU, but it also plays a major role in teaching on certain campuses in Australia and Canada and is gaining popularity in the moves towards distance education in the military (especially in the USA). One of the reasons for this widespread use is the ubiquity of the medium. In most developed countries every home is likely to have a TV set, so most people have experience of watching general service broadcasts and there will be some carry over of this experience to educational TV.

Many uses of TV in education have been developed. Television in some circumstances is simply used as a way of delivering a lecture to distant students. It is often seen as one solution to the problem of making lectures more efficient by multiplying the number of students who can be reached by a single lecturer. This is particularly the case in the USA where 'traditional' universities televise their lectures for transmission to a wider home-based student body, sometimes using cable TV. Television is also used in this way in multi-site institutions so that students at remote sites can receive a lecture synchronously with students in the lecture room; this provision may be supplemented by a local tutor available to answer questions or a telephone link for direct contact with the lecturer. Laurillard (1993, pp. 115, 116) refers to two different roles for TV – one to provide 'vicarious perception . . . [which] . . . acts as a solution to the problem of enabling large numbers of students to experience [an] aspect of the world directly'. She means, for example, a vicarious experience of fieldwork such as climbing a volcano and inspecting samples. However she values more highly her second category of 'vicarious conception', by which she means 'the way the teacher thinks about the topic'. She advocates the use of television to persuade the viewer of a line of argument or a way of seeing the world. She points out that various technical devices such as zooming, panning or montage can help to manipulate the viewer's experience of the world shown on film by mimicking such psychological features as selective attention, shifting attention or encouraging the association of ideas.

There has been a tradition of research into learning from media using comparative studies which aim to discover, for example, whether students learn more from a live lecture or a televised lecture (see Clark, 1983, for a review). Some studies even put forward the hypothesis that there is no difference in educational effectiveness between media, while other researchers argue that their task is to identify the unique features of each medium (see *Educational Technology Research and Development*, 1994). Much of the research in educational TV deals with students' ability to recall particular facts from programmes, but there are other features that should also be considered. As Bates (1982) points out, in a review of the literature on learning from educational television, the nature of that learning includes at least two aspects:

> a) the presentation of knowledge in a different way from the presentation of the same knowledge through another medium, thus providing a broader base of knowing – knowing *what* in different ways
> b) the development of certain mental skills in using knowledge – knowing *how* in different ways.
>
> (Bates, 1982, p. 34)

This second aspect is one that Salomon (1979) and Olson and Bruner (1974) have referred to and researched. The latter point out that

> to untangle the educational effects of these symbolic systems we again have to differentiate the knowledge of the world conveyed through the system from the skills involved in the mastery of the structure of the medium itself.
>
> (Olson and Bruner, 1974, p. 149)

One focus of research should be on the ways in which students develop the skills they need in order to learn from TV.

Usually a mixture of quantitative and qualitative evidence is used in evaluating the effectiveness of broadcasts (Summer, 1991). Case study methods can be used to look at individual programmes and questionnaire methods can be used to establish information about the role that educational TV can play and is expected to play (see e.g. Moses and Croll, 1991).

Broadcast TV has the disadvantage of fixed scheduling. Also programmes have to be viewed at one sitting (with perhaps one repeat showing) and are in that sense ephemeral. This means that it can be difficult for students to retain much of the detail from a TV presentation. The fact that most TV programmes are watched purely for entertainment compounds the problem: most viewers do not develop study habits suitable for learning from broadcast media, but tend simply to watch (or listen) passively. Indeed one popular line of research into the effectiveness of educational television stresses the importance of the psychological state of the viewer (see e.g. Gunter, 1985).

One disadvantage of all TV or video material for educational use is that it is time consuming and expensive to produce compared with text. Also, in the case of broadcast TV, which necessarily involves professional producers and technical staff, the academics inevitably lose some control of the programmes in either content, structure or teaching approach. Differences of opinion about any of these matters can lead to conflicts of interest between the academics and the TV professionals. Perry commented on the effect of this in the UKOU's very first year. He referred as follows to the tension between the perceived importance of the production quality of the programmes and pedagogical effectiveness:

> It was very hard for those of us without experience of television to

accept that it would take a whole day – sometimes two whole days – in a studio to produce a programme lasting only twenty five minutes. It was sometimes hard for the BBC to realise that the academic who was working flat out to prepare a correspondence text could not spare the time to work on the precise details of a broadcast related to that text until the text itself has taken on a reasonably final form. It was equally difficult for some of our staff to accept that the nice things they would like to show on television would be vastly too expensive and time consuming to create on videotape. Finally, it was sometimes difficult for the BBC production staff to realise that an attractive visual presentation of a particular technique would actually carry very little pedagogic weight when shown to students.

(Perry, 1976, pp. 82–3)

Over the years an accommodation has been reached and a good working relationship developed between the UKOU and the BBC. Meanwhile research has continued in an attempt to discover what factors influence the pedagogic effectiveness of programmes (see e.g. Gallagher, 1977 and 1978).

Videocassettes

The possibility of producing and delivering visual material on videocassette at first seems just the ideal solution to the problems posed for students by the use of broadcast TV – while retaining almost all the benefits of TV transmission. Two significant advantages of videocassettes over broadcast TV are that the student gains more control of the medium and that the length of session is no longer tied to the length of the transmission slot. However, an even more important aspect of videocassettes from a pedagogic viewpoint is the way in which their ease of stopping and starting can be used to provide cues for students' active engagement with the material. Thus, for example, students can be provided with information in one sequence, or asked to read data from the screen, and then directed to stop the tape in order to perform an exercise based on the information or data. Discussion of the results of the exercise can then follow in a later sequence. Crooks and Kirkwood (1988) have put forward three categories in which videos are potentially more effective than TV:

1. in providing vicarious experience.
   Students can be provided with the 'I was there' feeling because the pause and replay facilities of video allow them actually to make observations for themselves, at a leisured pace similar to that they could adopt were they actually to visit the environment concerned. Video formats also allow access to a number of different viewpoints, and can present the experiences of a number of participants (e.g. in case studies), material that might be too confusing within the span of a broadcast TV programme.
2. in imparting visually or conceptually dense information.

Effects such as moving pictures, rotating diagrams or models, super-imposing visuals, constructing parts of complicated diagrams piecemeal, illustrating processes using animations, whether computer generated or not, can be achieved easily using videos. Although all these devices are available for broadcast TV too, the great benefit of a videocassette delivery system is that segments can be replayed to watch for different aspects of these processes. So video is a potentially invaluable tool for dealing with visually dense material. Since conceptually dense material will need a variety of approaches for students, video is of course just one part of the armoury of techniques an instructor could try.

3. in triggering reflection or group discussion.

Video presentations can go some way towards providing relevant vicarious *joint* experience to be the subject of group discussion. This would help to deal with the problem raised in Chapter 3 that distance students need common experiences to share. One way in which reflection can be triggered is by the use of case studies. Two examples of multiple media case study approaches will be described in Section 5.

One problem with videocassettes containing many short sequences is that it can be difficult to locate a particular sequence.

> Although video-cassettes make it possible for student to stop the tape . . . the mechanics can make it difficult to exert that control in a precise way. A caption at the beginning of each new sequence or a sequence number at the top of the screen may be sufficient for students to locate a sequence at a gross level . . . Some OU video sequences have used clock or page numbers, both of which are particularly useful for relocating a section for reviewing . . . When students are studying a particular sequence, the clock or 'page number' enables them to gauge their progress and to assess how much material remains before it ends.
>
> (Crooks and Kirkwood, 1988, p. 141)

As well as screen clocks or page numbers, or other similar devices that simply help students to navigate round the video, it is easy to introduce mechanisms to encourage students' active engagement with the material. For example, captions announcing tape stops can be used to cue points at which students should pause the video in order to consider various options, think about particular questions (which may be posed on the tape itself or appear in the associated printed notes), perform data analysis or carry out calculations.

Radio

Although radio may seem in some respects an old fashioned medium, it has a number of features which may be creatively exploited in the educational context. Archive material can bring historical material to prominence: to hear

Einstein's voice, or to hear the original news report announcing a scientific discovery, can be a thrilling experience. There is obviously enormous potential in the use of radio for students who have impaired vision, or who have problems reading. There is the possibility of broadcasting an academic debate or a tutorial which students can at least listen to, if not take part in. Radio programmes can be recorded and transmitted quickly, so can be very useful in delivering up-to-the-minute programmes linking topical issues and events to ideas, principles and theories presented in text material. There are even a few examples of educational radio phone-ins, in particular the Radio Tayside experiment (Auchterlonie, 1989), though the potential of these does not seem to have been fully developed.

Audiovision

Radio broadcasts share some of the disadvantages of broadcast TV in that they are ephemeral, and that some of the potential of a friendly voice leading the student through difficult topics is not totally fulfilled if the student cannot stop and listen again to a particularly difficult passage, or be referred to useful visual cues. A partial solution is 'radiovision', which involves the advance provision of visuals to which a broadcast radio programme will relate. However, the power of audiovision as a teaching and learning tool is most fully realized when visuals are harnessed to audiosequences packaged on cassette. Students can then start, stop and replay the tape at will, and be 'talked through' a lengthy or complicated activity. As with videocassettes, it is in the integration of audiosequences with other course components demanding active student participation that the real teaching power of the medium lies.

3. AUDIOVISUAL MEDIA IN DISTANCE EDUCATION

As outlined in the previous section, audiovisual media can play a valuable role in any multiple media educational package. For distance teaching institutions, however, audiovisual media have certain very particular advantages and drawbacks. Some of these are illustrated in Subsection 3.1, which presents a brief history of the development of audiovisual components for science courses at the UKOU. Other important issues arising from the use of audiovisual media in distance education packages are highlighted in Subsection 3.2.

3.1. The use of audiovisual media at the UKOU

In the first decade of the UKOU's operation, during the 1970s, the institution's audiovisual activity was almost exclusively based on broadcast TV and radio. By the mid-1970s, the University was transmitting about 300 television and 300 radio programmes each year. At first around 10 per cent of students could not receive TV broadcasts and different parts of the University responded to this fact in different ways. Students were actually advised not to register for science courses unless they were able to receive broadcasts, as these formed part of the Science Faculty's overall strategy for providing practical experience. The other three original faculties (Arts, Social Sciences and Mathematics) designed their TV programmes to allow students to complete the courses even if they were unable to receive broadcasts. In general, the TV and radio programmes were closely and explicitly linked with other components of the teaching system, especially in the Science Faculty. Perhaps as a result of this policy, viewing and rating figures were higher for science courses than for many others (McIntosh, 1974). It was soon found that some of the problems with this type of provision arose simply from the inflexibility of the transmission schedules. It had been assumed that students would use a weekly or fortnightly broadcast to help pace and organize their study and, in fact, many early plans expected that students would have reached a certain point in their study before watching a particular programme. However, it turned out that programmes had little pacing effect and in many cases students were two or three weeks behind in their study when they watched the week's broadcast. As will be discussed in Chapter 11, it was found that the pacing function originally expected of the broadcasts was in fact largely taken over by the assignment schedule imposed by the continuous assessment policy.

It was thought that radio programmes, which were cheaper to use and could be made quickly at short notice, would be particularly useful for emergency or remedial transmissions, and some use was made of the facility for 'stop press' announcements or more substantial remedial programmes, e.g. a series of radio feedback programmes produced for an organic chemistry course. In general though, radio was less popular with science, mathematics and technology students (although very popular in arts courses). In the Science Faculty this led to an almost complete switch from radio to radiovision and audiocassettes. Indeed, the amount of transmission time allocated for radio fell substantially in courses across the University and by the end of the 1980s, Bates (1990) found that audiocassettes were the most widely used technology in the UKOU after print, with 750,000 hours of material on cassette.

By the early 1980s, although most TV programmes were still broadcast, videocassettes were beginning to have a significant impact on teaching styles. The most immediately attractive features of videos were their flexibility: it became apparent that many students found the transmission times of broadcast TV inconvenient, and as home ownership of videorecorders grew there was anyway an increasing tendency for students to record UKOU programmes for later viewing rather than watching them 'live.' While a substantial number of students still did not have easy access to videoplayback facilities, the University was reluctant to make videocassettes an extensive element in much of its course provision. However, the popularity of the video loan scheme whereby programmes scheduled for TV transmission were made available to students on cassette, and the rapid increase in home ownership of videorecorders (access among UKOU students was estimated to be 90 per cent by 1991), meant that by the beginning of the 1990s the issue of access was no longer considered to be a problem. Course teams were free to put essential material on to video if that was the best medium for presentation of that particular element of their instructional strategy.

It is instructive to realize that this shift away from broadcasting was no surprise. One of the University's first official publications, the 1971 prospectus, contained the following paragraph reproduced from a 1969 speech by the first Chancellor, Lord Crowther:

> We start in dependence on, and in grateful partnership with, the BBC. But already the development of technology is marching on, and I predict that before long actual broadcasting will form only a small part of the University's output. The world is caught in a communications revolution. Every new form of human communication will be examined to see how it can be used to raise and broaden the level of human understanding.
> (Crowther, 1971)

The Broadcasting Bill which came into force in 1990 made available more channels through terrestial and satellite transmissions and developments in cable networks. This has resulted in a greater competition for audiences, especially at popular viewing times. In fact transmission times made available to the UKOU by the BBC have been gradually decreasing, a decline partly accelerated by the increased use of videocassettes. Even for broadcast programmes, the volume of home recording accounts for all but 8 per cent of the total viewing of UKOU television (according to Kirkwood, 1990). However, it is important not to confuse issues about the means by which teaching material is distributed with educational issues, a point we will explore in more detail in Subsection 3.2.

3.2. Institutional and educational issues

The distinction between format and method of distribution is an important one for distance education institutions. Figure 7.1 illustrates the various possible combinations of televisual media.

Quadrant 1 corresponds to the standard TV programme that is broadcast just once or twice during each presentation of a course and is designed to fit a particular length of transmission slot. (Most of the early TV production of the UKOU fell into this category.) Production costs for such programmes are considerable, not least because television companies impose high standards for broadcast material. Distribution costs depend on charges for transmission time, but are economic on a per capita basis only for courses with a fairly large student populations. However, a national broadcasting presence does have a 'shop window' effect for a distance teaching institution, helping to spread awareness of the courses beyond the registered student body.

For example, the UKOU has always attracted a large general audience for its TV programmes. This finding emerged from the fact that its programmes sometimes appear in the BBC's Broadcasting Audience Research Bureau (BARB) ratings for which the minimum audience for inclusion in the statistics is 500,000 viewers. Radcliffe (1990) refers to audiences for UKOU TV programmes regularly reaching 200,000. On some courses therefore these eavesdroppers or 'drop-in' viewers can be assumed to outnumber the registered student viewers by many more than ten to one, and around half of these (Taylor, 1992) were found to engage in some follow-up activity.

Quadrant 2 in Figure 7.1 represents the situation in which a programme is made to broadcast standards and length, but is actually distributed to students on videocassette. Clearly this is unlikely to be a wholly satisfactory state of affairs, since it does not fully exploit the strengths of either medium. It may be forced upon the institution in the case of low population and/or high level courses, for which it is difficult to justify the costs of transmission time when reproduction and mailing of tapes is clearly a cheaper option (see Bates (1987) and Curran (1990) for further analysis of costing implications). There can be a positive side to this compromise, however, in that it may prolong the usable life of programmes which for stylistic reasons the broadcasting companies no longer wish to transmit but which still contain valid instructional material from which students can benefit.

Quadrant 3 in Figure 7.1 corresponds to material specially designed to be distributed on videocassette. This is educationally the preferred option, since in addition to the other advantages of video presentation already rehearsed in

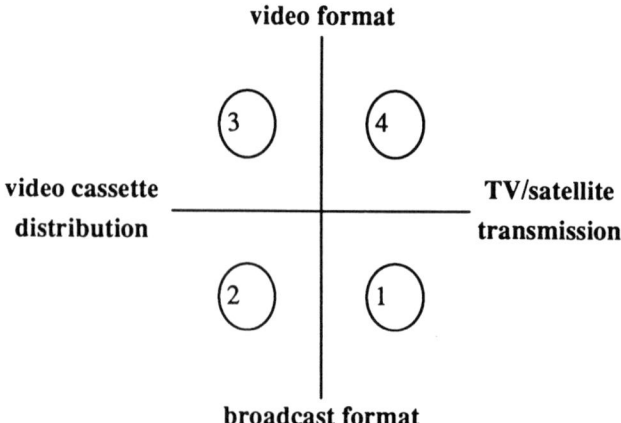

Figure 7.1. Formats and distribution methods for televisual components of distance education courses.

Subsection 3.1, the lengths of the video segments can be tailored to the teaching and learning strategies, instead of being dictated by the length and availability of transmission slots.

Quadrant 4 in Figure 7.1 represents an alternative distribution strategy, whereby material in video format is transmitted, usually at very antisocial hours, for students to record off air. This method of distribution has been relatively little exploited to date, at least in the UK, but may offer interesting developments in the future.

Once the format and means of distribution of televisual (or indeed radio) material has been decided, the next issue to be considered is the provision of associated 'orientation' devices for the students – for example suggestions for ways to become involved in the programmes or for structured follow-up activities (Convery, 1990). These suggestions usually come in print. Certainly in the UKOU, very few programmes or tapes are left to carry the whole teaching burden without some text accompaniment. These pieces of text can be vitally important in overcoming the ephemeral nature of much broadcast television and allow the instructional designer to set assignments on the programmes. Such notes can be provided for reading both before and after viewing the programme. Pre-broadcast notes can help orient students and warn them of significant points to look out for in the broadcast. Post-broadcast notes can provide reminders of key visuals from the programme and suggest exercises which can be used to consolidate the main teaching points in the programme. Figure 7.2 is an example of a set of broadcast notes showing both of these features.

Pre-broadcast notes

This programme is related to Objectives 1, 2 and 11–13 of Block 4, and is designed to enhance your understanding of these Objectives. It summarizes the surface and internal features of the four inner planets—Mercury, Venus, Earth and Mars—and the Moon, and uses these to comment on past and present internal processes during the geological evolution of these planetary bodies. The surface and internal features of the planetary bodies are reviewed in the programme and are summarized in Table 1 and the Post-broadcast notes.

Post-broadcast notes

With the exception of the Earth, the inner planets appear to show no surface tectonic activity. On the Earth, the contrast between geologically young oceans (<200 Ma in age) and old continents (up to 3700 Ma in age) indicates the occurrence of some form of internal convection. The surface expression of this convection is plate tectonics, and measurements of the rates of instantaneous relative motion (Block 4, Section 2.5) can be made by satellite.

Plate motion is an expression of the Earth's heat loss (Block 4, Sections 3.4–3.7). The heat flow from the Earth is generally within the range 40–120 mW m^{-2} and averages $c.$ 80 mW m^{-2}. Exceptionally high values are measured on ocean ridges (average $c.$ 150 mW m^{-2}) and in areas of young continental volcanism (note that the profile shown in the programme is schematic; you should refer to Block 4, Figures 29 and 43 and to Tables 6 and 7 for exact heat flow values). Because extrapolation to depth of temperature gradients measured near the surface leads to implausibly high estimates of mantle temperature, the temperature gradient must *decrease* at depth, within the mantle. There is a gradual decrease in temperature gradient within the crust, and also at a greater depth, which is assumed to represent the base of the lithosphere; heat flow is predominantly by *conduction* within the lithosphere, but is assumed to be predominantly by *convection* in the asthenosphere below the lithosphere. Because convection is a much more effective mechanism of heat transfer than conduction, the temperature gradient within the asthenosphere is less than in the lithosphere.

Figure 7.2. An example of printed 'Broadcast notes' which accompany a TV programme on a course on the *Earth's Structure, Composition and Evolution* (Open University, S237, TV 08 notes).

Accompanying notes can be equally important for video material, but given the controllability of the medium their purpose need be less directed towards orientation and summary. Instead they can be focused more on promoting active engagement with the video sequences, for example setting exercises based on information presented on the video. (Good quality reproductions of tables, e.g. data on graphs, can usefully be included in such notes, to avoid the tape damage which might occur if students had to work for long periods from a paused frame.)

In seeking to encourage active learning from video material, another important issue for instructional designers concerns the length of the segments students are expected to watch. In the distance education context, it has to be recognized that students may face considerable competition from other members of the household for use of the videorecorder. Sequences therefore have to be of sufficient length and importance for students to feel it is worth working with them at the designated point in the course. Segments that are too short (e.g. last only a few minutes) are usually perceived as not worth bothering with on their own; segments that are too long (e.g. in excess of 20 minutes) may encourage passive viewing in the TV entertainment mode already alluded to in Section 2. Varying the length of sequences during the course of a programme is claimed to be useful in motivating students (Bates, 1985).

The use of any instructional device, especially one as expensive as televisual media, must of course be justified in terms of teaching effectiveness. The UKOU has always undertaken detailed evaluation of the use of audiovisual components in its courses, and overall results show that students value and learn from the media mix. Survey statistics have shown that those who watch programmes the most obtain the highest grades (Bates, 1975), borderline students rate TV programmes as most helpful (Gallagher, 1977) and the more TV programmes there are on a course the lower the drop out rate (Parlett and Woodley, 1983). The use of audiocassettes has been evaluated in a number of courses by Durbridge (1980, 1983, 1984) and for all courses as part of the annual survey of new courses (e.g. Lawless, 1989); both report a high degree of student enthusiasm for audio and audiovisual sequences in courses when they have been properly integrated. In general, Science Faculty TV programmes have been consistently watched by a very large proportion of students and have been rated very highly in terms of their helpfulness (Kirkwood, 1991). The switch to videocassettes has been monitored and case study evaluations of a number of courses conducted (Crooks and Kirkwood, 1990) which suggest that videocassettes are generally well received by students.

UKOU programmes also receive a surprising amount of positive comment from unexpected quarters. One celebrated TV critic has compared Open University broadcasts with general broadcasting in very positive terms. Even the radio programmes which have struggled to find an appropriate role in course structures have been praised.

4. AUDIOVISUAL MEDIA FOR SCIENCE STUDENTS: EXPERIMENTS, INDUSTRIAL CASE STUDIES AND VISUALIZATION

TV and video

Audiovisual media have a very important and particular role to play in the study of science at a distance. TV and video are particularly useful for illustrating:

1. experiments, especially if expensive, difficult, risky for students to perform or unethical if done in bulk (e.g. dissections);
2. field work or experiences otherwise inaccessible to students;
3. motion (either in slow-motion with superimposed graphics for analysis of complex motions or speeded up, e.g. time lapse photography of plant growth or deposition);
4. 3D effects;
5. animations;
6. affective material (e.g. interviews with famous scientists);
7. information about science in a social context.

Categories 1 and 2 are of particular relevance to remote students. As mentioned in Chapter 2, the difficulty of providing an experience of practical work is an important issue for distance education in science. The advantages of TV or video for this purpose include the fact that in a studio setting carefully planned and rehearsed experiments make it possible to show students clear results. Furthermore by combining cameras with other instruments such as microscopes, remote students can make observations and take part in experiments, at least in Laurillard's 'vicarious' sense. Haight and Jones (1987) discuss how to overcome some difficulties presented by the use of clock reactions with large numbers of students by using videotapes (without narration) in stages so that observations can be recorded. For direct teaching of practical skills, it is not surprising that commercial production of instructional videocassettes now includes the possibility of learning about analytical instrument maintenance and use of instruments like gas chromatographs and mass spectrometers (see e.g. Maugh, 1985).

Some experiments are most easily observed by taking a film crew to a university or industrial laboratory and using their facilities. This is only one example of the way in which television can take students into situations which provide great learning opportunities. Other examples include vicarious field trips in geology and biology, which are for remote students a valuable substitute for time-consuming visits to sites they would in any case find it difficult or impossible to reach. As already noted, video presentations allow students actively to participate in 'virtual experiments' or field trips, by carrying out observations or analyses for themselves from information supplied on the tape.

Categories 3–5 are televisual techniques that can be used to help explain and clarify difficult ideas using different representations. Two-dimensional representations such as graphics or pictures, or animations, computer generated or otherwise, can be very useful aids, as can 3D representations such as realistic or schematic models. The ability to show the rotation of 3D models is particularly useful. One obvious use of televisual media here is in the teaching of stereochemistry which deals with the 3D arrangement of atoms in space, and the ways in which equilibria and reaction rates are affected by changes in the 3D relationships of atoms during a reaction.

Categories 6 and 7 refer to the possibilities of illustrating the applicability of scientific ideas in real life situations, and the potential for increasing student motivation. The ability to combine information about scientific phenomena with evidence about the use of science in a particular environment is especially valuable. The use of television to 'enhance the message' by including 'talking heads' – whether famous or just expert – can play a part in increasing motivation. Using TV to provide context and motivation is an approach seen most clearly in audiovisual material used for case studies. This case study approach can be particularly effective in technology to present industrially related material, but has also been used very successfully in mathematics and science education where edited film or videorecording of actual situations is presented to students to interpret or explain on the basis of material covered elsewhere in the course. We will illustrate such a case study approach in Section 5.

Audiocassettes

As briefly mentioned in Section 2, audiocassettes are most effectively used when coupled with other material in audiovision mode, in such a way that the taped commentary helps students interact with that material. A friendly voice 'talking through' difficult calculations or arguments can be of considerable encouragement to students, and survey evidence collected at the UKOU shows that students greatly appreciate this kind of tutorial aspect of audiovisual sequences. A typical student comment is that it is like 'having the professor in the room talking over my shoulder' (Durbridge, 1983). This kind of positive reaction has reinforced the policy decision made by some UKOU course teams to use audiovisual sequences to teach the most difficult material in a particular course. In physics, for example, audio commentary integrated with print has been found to be a reassuring and effective way of guiding students through complex formulae, calculations and graphs, and of helping them to develop problem-solving skills. Audiocassettes can also be very useful in conjunction with experiments or samples, as students can listen to the commentary while looking through a hand lens or microscope, or while carrying out practical work: 'for stable images, objects, and controllable actions, the audiovision combination can be a very effective educational medium' (Laurillard, 1993, p. 113).

We will give examples of the integration of audiocassettes with other course components in Section 5.

5. AUDIOVISUAL MEDIA IN USE

This section documents some of the uses of audiovisual media in science learning at the UKOU, including highly integrated video for dealing with conceptually dense material, video led courses, video for providing vicarious experience of experiments, audiocassettes for problem solving and for practical work and multiple media case studies. This list of examples is in no way exhaustive; it is simply intended to give a flavour of the many ways that audiovisual media can enrich students' experience of science learning.

Highly integrated video

A second level physics course *Discovering Physics* makes extensive use of videocassettes for dealing with difficult parts of the course. One feature of the instructional design in this course was a decision to make the video interactive by asking students to stop the tape at certain points and answer questions which require observations taken from the video and sometimes calculations using these observations. Typically, presenters take the students through a sequence of arguments and demonstrations, cued by a series of questions which the students deal with after each subsection of the video. An example of the text accompaniment is shown in Figure 7.3.

Video led instruction

A third level Earth science course *Sedimentary Processes and Basin Analysis* contains a number of video sequences (six videocassettes involving fourteen hours of viewing) which substitute for field trips. The study commentary which accompanies the video directs the student to read relevant sections of a set book.

Audiocassettes

In another geology course, an audiocassette on *Mineral Structures* requires students to build structures from balls and spokes provided as part of a home experiment kit. Figure 7.4 shows the printed notes that tell students what material they need to assemble before listening to the tape.

A different use for audiocassettes, but one that has been especially popular

The video sequence for this Unit looks in more detail at the operation of dust precipitators, and also provides an introduction to the printers described in the next subsection. The sequence is located on videocassette 2 band B.

▼ *NOW START THE TAPE* ▼

■ *TAPE STOP 1* ■

> **ITQ 8**   Suppose a dust particle is shown in the region between the plates, closer to the negatively charged plate, that it carries 10^6 electrons and that the electrostatic force on it is of magnitude $2 \times 10^{-7} \mathrm{N}$.
> (a)   In which direction will the particle move?
> (b)   Would the magnitude of the force on the particle be greater, smaller, or much the same if the particle were in a position equidistant from the two plates?
> (c)   By roughly what factor could the magnitude of the electric field between the plates be increased before breakdown would be likely? (*Note*: as stated above, the magnitude of the field required for breakdown in air is about $3 \times 10^6 \mathrm{N\,C^{-1}}$.)

▼ *NOW RESTART THE TAPE* ▼

■ *TAPE STOP 2* ■

> **ITQ 9**   Suppose the dust particle described in ITQ 8 has a mass of $6 \times 10^{-11} \mathrm{kg}$.
> (a)   Calculate the magnitude of the gravitational force on the particle. How does this compare with the electrostatic force on it? Would the precipitator work upside-down?
> (b)   Taking the mass of a typical atom in the dust particle to be $6 \times 10^{-26} \mathrm{kg}$, estimate the ratio of the number of excess electrons to the number of atoms. By how much could this ratio be reduced before the precipitator ceased to operate?

▼ *NOW RESTART THE TAPE* ▼

Figure 7.3. An example of print accompanying a video sequence which shows a number of questions which students need to tackle while working with the video (Open University, S 271, Unit 8, p. 34).

---

with physics course teams, is to tie a spoken piece of teaching to a series of specially designed 'Frames'. An example of this approach is shown in Figure 7.5, which illustrates part of an audiovisual sequence on diatomic molecules from a second level course *The Physics of Matter*.

This figure exemplifies two of the strengths of this kind of mixed-media presentation. In the commentary accompanying Frame 2, the students are talked through the steps required to find the vibrational frequency of the H^+ ion. A more discursive approach is possible than would be the case with a linear text presentation, and the frame also illustrates for students a kind of layout for notes that they might find useful in other areas of their work. In Frame 3, students are required to read information from the graph and use it to calculate various quantities. The commentary sets out the task for them, instructs them to pause the tape while carrying out the calculations, and gives additional help in arriving at the answers. Many students find this a reassuring and confidence-building way of approaching new or difficult material.

For this audiovision sequence you will need cassette AC 246 (Band 1). Block 1, Colour Plates 1–34, the balls and spokes in your Home Experiment Kit and, if possible, some Blu-tack. Note that reference will be made to Block 1 Figures, Table 7 and the Colour Plates in the following order:

Figures 19 and 20; Table 7; Figure 21; Plates 7, 12, 13; Figure 28; Plates 8 and 9; Figures 29 and 30; Plates 14, 15 and 10; Figures 31, 32 and 35; Plates 17–21.

We suggest that you insert slips of paper in the Block at the appropriate pages.

The sequence starts on the tape by revising the concept of co-ordination number in mineral structures. After about five minutes you will be asked to stop the tape and complete Exercise 1 (below). The tape then continues for about 30 minutes by asking you to study pictures and to make models of the structures of silicate polytetrahedra. When you come to do this, be careful to note that, in each model (as in *CB*, Plates 7–11), silicon is represented by *black* balls (rather than red as in Block 1) and that each silicon ball should be joined to four oxygen balls (red) using the spokes. In *every* model, three of the oxygen balls attached to each silicon ball should rest flat on the table with the fourth pointing vertically upwards.

When you have everything ready you should start listening to the tape.

Exercise 1

Using the plastic balls provided — any colour will do for this exercise — try to make the three models shown in Figure 1 which involve (a) tetrahedral, (b) octahedral and (c) cubic arrangements of balls. You will need some means of holding the balls together to make models (b) and (c) — this is where the Blu-tack will help.

When you have made the models:

1 Rank them according to how large a ball you judge could be fitted into the hole enclosed by the balls.

2 Decide what co-ordination number such a central atom would have in each model.

When you have done this, switch on the tape again.

At the end of the sequence, return to the Main Text of Block 1 (p. 50) and do ITQ 16, which summarizes the main points we hope you will have gained from this tape. The details of the appearance and compositions of silicate minerals which were introduced on the tape are discussed in more detail in Section 3.3.

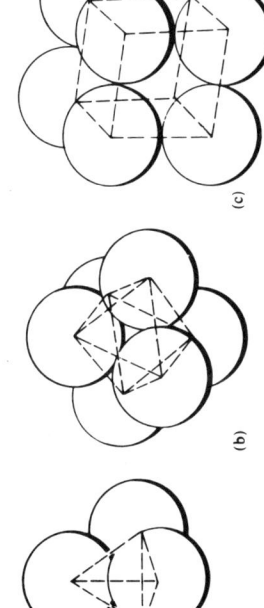

(a) (b) (c)

FIGURE 1 For use with AV 01, Exercise 1.

Figure 7.4. An example of printed instructions to accompany an audiocassette on *Mineral Structures.*

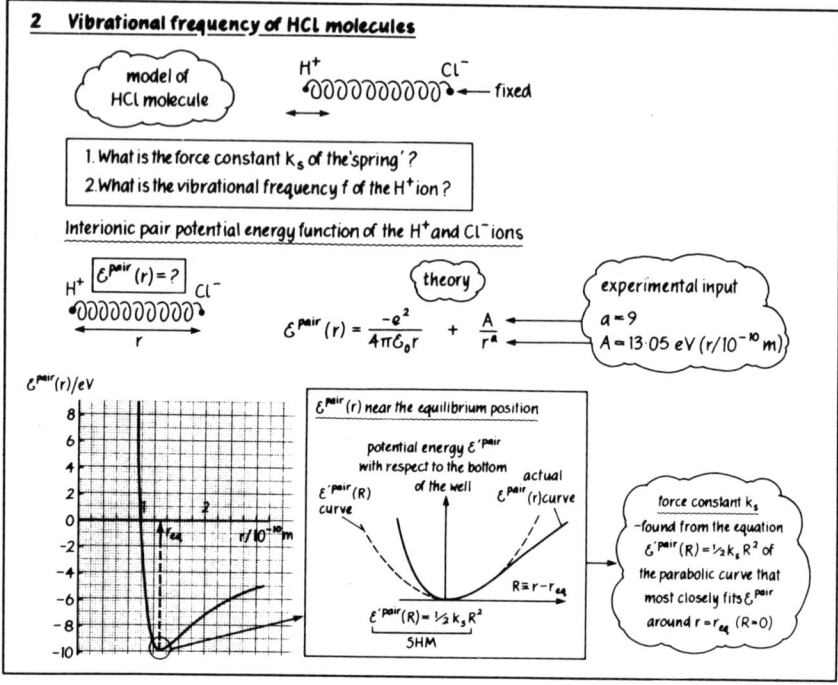

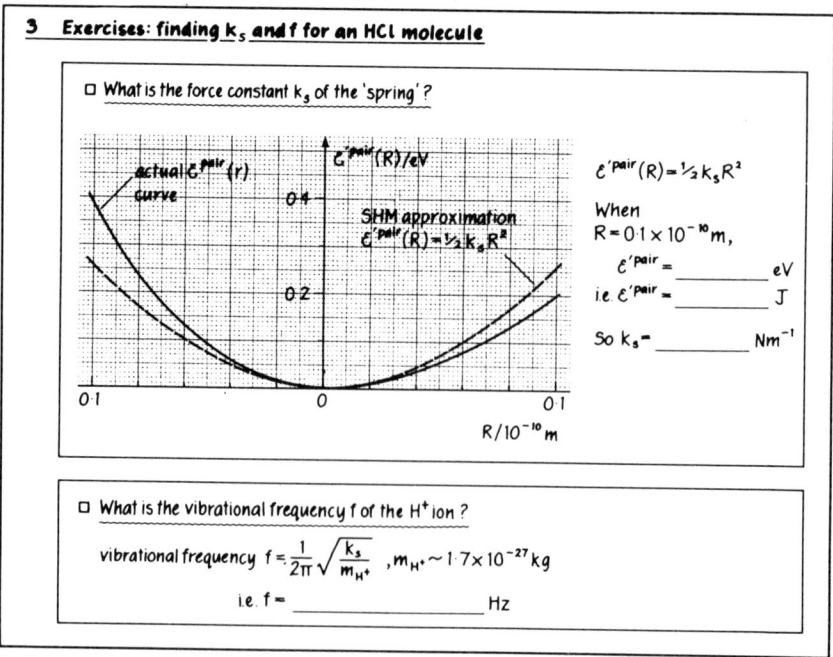

Figure 7.5. An example of 'tape frames' to accompany an audiocassette sequence on the classical description of a diatomic gas (Open University, S272, Unit 3, pp. 15–16).

Multiple media case studies

An even more broadly based 'vicarious perception' can be provided by *combining* video and audio data to present a rounded case study description that not only provides experience but a description of that experience. Such a mixed medium presentation is used in a course for teachers on *Science in the Primary Curriculum*. The course is led by both video and audiotapes of children doing, and talking about, science, providing a variety of stimuli such as observations of young children at play and scenes from news, current affairs and documentary programmes. These examples are designed to encourage reflection on the nature of science, the ways in which attitudes and images of science evolve, and the influence of these ideas on both the teachers and the children they teach. The media mix also tells a story that unfolds and is embellished in stages. Each stage is supported by video and scenarios, with associated audio extracts to add depth. An audiocassette provides examples of childrens' talk about science and their ideas and questions. These examples are used to explore with teachers how to use discussion in class groups and small groups to assess children's learning, and how to question children effectively. Multiple media courses are the best way to achieve the necessary blend of theory and practice in distance education courses for science teachers, illustrating and analysing what other teachers are actually doing in the classroom and why they choose to work in the ways that they do. The role of audio and video is to provide starting points, illustration and evidence, as well as inspiration and illumination.

The use of video to support case study material has also been used extensively in chemistry courses. A second level course on *Inorganic Chemistry* includes several case studies, presented in text format but heavily reinforced and illustrated by video sequences. Figure 7.6 illustrates the interfacing between these components.

6. CONCLUSIONS AND FUTURE DEVELOPMENTS

Audiovisual media can help considerably in rounding out the educational experience that it is possible to provide for remote students. In science courses, such media can be especially valuable in giving students exposure to field and laboratory environments, and in granting them 'access' to industrial facilities. Other features of audiovisual media can also be exploited to give students more sense of interaction than is usually possible in a textual presentation.

Monitoring and analysis of the educational use of audiovisual media, at the UKOU and elsewhere, has shown that there can be a world of difference between academics' and producers' expectations of how audiovisual media will be used, and the ways in which students actually work. The intention of the course designers may be for students to stop a cassette in order to complete a particular piece of work and then to switch on again. But packaging the

The video sequence accompanying CS 2 lasts around 18 minutes and supports the text of Case Study 2 up to Section 8. The topics dealt with are indicated in the table below.

Time	Related CS 2 Sec.	Topic
0–2 min	2	demonstration of (thermal) semiconductivity
2–5 min	1	band theory; development of energy 'band' from individual atomic energy levels as continuous solid builds up
5–7 min	2.1	simulation of metallic conductivity on the scale of electrons and atoms
7–10 min	3	simulation of semiconductivity
10–12 min	4	simulation of photoconductivity
12–14 min	5	simulation of impurity semiconductivity
14–16 min	6 and 7	simulation of processes operating in the electron-poor/electron-rich junction
16–17 min	6, 7 and 9	demonstration of photovoltaic cell

From Section 4 on (that is, after about 12 minutes) use is made of a split screen to illustrate simultaneous processes. You will probably need to watch the split screen sequences several times to make sure you have understood:

(i)   what is happening in the lattice;

(ii)   what is happening to the energy of the bands;

(iii) how these processes fit together

The crucial part of the sequence is the last 5 minutes, where we consider impurity semiconductors and their combination into a junction. This is where you most need to concentrate your attention and viewing time, and it is important here that you have understood the preceding material. During this final five minutes, you will need to concentrate on processes taking place (in different parts of the screen) which affect:

bonds in the lattice

development of electric charge on some atoms

migration of electrons

gradient of the 'bending' of energy bands in the neighbourhood of the junction

Figure 7.6. Orientation notes for a videocassette that is integrated with textual material into a case study. This particular case study is focused on the operation of solar cells, but is also the vehicle for teaching about bonding in solids and about the properties of semiconductors (Open University, S247, VN1, sequence 8).

material on cassette gives the student ultimate control over the on/off and pause buttons. Detailed evaluation is therefore necessary to find out exactly how students do use the audiovisual components of course material.

In conjunction with progress in understanding the educational uses of audiovisual media, developments in the technology need to be considered. Here, there are both problems and opportunities. The shift to the provision of videocassette and audiocassettes has been accompanied at the UKOU by a raised perception of the value of these media for learning. The fact that they are more under the students' control than broadcast TV contributes to their educational effectiveness. However, there are costs to the institution's profile with the general public in a switch from broadcast TV and radio, in that the 'shopwindow' aspect of the transmissions is lost. As Sargant points out

> Much educational material that is now broadcast is *de facto* 'narrow-cast' and aimed at specific and known target groups ... As new options become available it is increasingly sensible and necessary for us to be selective and to use terrestial broadcasting only for content and target groups which need it, and other delivery options (either satellite or terrestial ones) when they are equally or more appropriate.
>
> (Sargant, 1989, p. 61)

Satellite broadcasting has the potential to increase the amount of educational material available for drop-in viewers and serious students alike, but it remains to be seen whether this potential is ever realized. We will return in Chapter 15 to the implications of new technological developments for the future of distance learning.

8

COMPUTER MEDIATED LEARNING

Thinking about the use of computers in education does not mean thinking about computers. It means thinking about education.

(Darby, 1989)

1. INTRODUCTION

The rapid spread of computers into many aspects of life – business, research, and leisure activities, among others – was one of the major revolutions of the 1980s. The educational sector also participated in this revolution, in two rather different ways. One manifestation of the increasing role of computers is the expectation that many children will develop computer literacy as part of their basic education. At further and higher education levels, specialist computing courses are much in demand. The acquisition of computing knowledge and skills as an educational objective in its own right is, however, largely outside the scope of this book. Of greater interest in the present context are the many new fields that are continually being opened up by the use of computers as tools in the educational process. We refer in other chapters to the importance of computers in the overall administration and record-keeping systems of large distance education institutions. In this chapter we shall be mostly concerned with the use of computers as educational instruments, either as teaching tools (i.e. as an addition to or a substitute for some other form of instruction) or as aids to the learner. Computers can enhance learning in a large variety of ways; here we shall highlight aspects of special relevance to science students (most particularly in relation to practical work) and to those involved in distance education (for example the use of computers to increase opportunities for communication between isolated students).

Our global term to cover this wide range of examples is computer-mediated learning (CML). There has been considerable discussion in the literature of the distinctions between CAL, (computer-assisted learning/instruction) and computer based learning/training (CBL/T). However, as every author tends to invest each individual acronym with a subtly different shade of meaning, and moreover as usage varies on either side of the Atlantic, we shall avoid adding a further layer of definitions by simply sticking with the term CML and applying it in its broadest sense. In this chapter, we will confine the discussion to applications based only on computers, deferring consideration of multimedia presentations until Chapter 9.

2. THE COMPUTER AS AN EDUCATIONAL TOOL

Only a few years ago, the use of computers within multiple media teaching systems was limited to programmed instruction – typically of the drill-and-practice type. In this kind of role, computers can excel – provided they have been well programmed! Unlike human tutors, computers can repeat their response many times without losing patience and can always devote 'full attention' to each individual student user.

Drill-and-practice routines are now falling out of fashion, at least in the UK, but multiple choice, computer mediated questions continue to be popular, whether as part of formal assessment (see Chapter 11) or for diagnostic purposes. Diagnostic frameworks have also been expanded into more interactive systems, to provide a conversational type of tutorial program.

The benefits of structured programs for instruction are fairly obvious: students can use them as a means of self-assessment, obtaining feedback on their achievement of course objectives, diagnosis of their areas of weakness and targeted remedial help. Feedback serves mainly to correct misunderstanding on the student's part and can range from a simple yes/no response to substantial corrective information which can itself take the form of new instruction. If the author is sufficiently industrious the student interacting with such a program can well form the impression that he or she is communicating with an understanding teacher. However there can be negative aspects too. Often the program's next action is determined only by the student's last response – or at best by some statistics on the student's performance on earlier parts of the program. The possibility of the 'individualized' instruction thus constructed providing the best teaching sequence for each and every student is fairly limited. Another problem presented by these tutorial programs is the assumption that systematic presentation is more important than a student's own mode of learning. The programs concentrate on the acquisition of knowledge by the students rather than the application of individual learning styles.

There are other educational uses of computing besides the overtly instructional. Many of these are based on *experiential* strategies, with simulation, modelling, problem solving and games being important examples. There are areas in which the only safe or practicable way of offering a real learning experience – for example in discovering exactly what combinations of factors lead to a disastrous outcome – is to base it on a computer simulation. The data processing capability of the computer is a vital element here, allowing students to explore relatively quickly the response of a model system to a variety of input conditions. Of course, in a simulation the model has already been created by the programmer, and the student should be aware that he or she is using a simplification of reality set up by someone else. Often, this aspect is not important – it is observing the behaviour of the model that constitutes the learning experience. In other areas, it is the problem-solving or model building process itself that is crucial. With the increasing sophistication and user-friendliness of commercial software, students do not necessarily need programming skills to participate in this kind of active learning: imaginative use of a spreadsheet can take students a long way. Figure 8.1 shows an example of one class of problem that can be effectively tackled by using a spreadsheet; it is interesting that a compartmental problem of this type would more conventionally be handled by mathematical modelling involving differential equations. Good modelling exercises and simulations have some definite goal for which the students are aiming, and under these circumstances, the whole experience can be very motivating and even fun. Malone (1981) cites the presence of a goal as being one of the distinguishing features in successful educational computer games. The attributes of challenge, fantasy and curiosity, which he identifies as the most intrinsically motivating aspects of computer games, are also usually present to a greater or lesser extent.

Computers are also valued for what Kemmis *et al.* (1977) describe as their 'emancipatory' uses, that is their ability to be a labour saving device which removes drudgery from tedious tasks, their database capabilities which enable students to have quick access to information, and their calculation capacity which allows the rapid display and statistical analysis of data. This emancipatory value also could be attached to the communication capacity of computers, creating new ways for learners and teachers to interact; this aspect will be further explored in Subsection 3.2.

In considering the various sectors of CML, a variety of classification

---

PURGING A GAS STORAGE TANK

THE PROBLEM

A gas storage tank has a volume of 3000 m³. It currently contains methane. The tank must be emptied so that it can be cleaned and inspected. Safety regulations require that it should contain no more than 1 part in 100 of methane before work (which may include welding) can be started.

Nitrogen is available and can be pumped into an opening near one end of the tank. Another opening (near the other end) will let gases escape. How much nitrogen will you need to dilute the methane effectively?

Figure 8.1. A problem designed to introduce heuristics and the use of spread sheets as a tool for problem-solving (from *How to model it: problem solving for the computer age* by Starfield *et al.*, 1990, p. 3).

schemes have been proposed. Rushby (1979), for example, used a four paradigm set, in which simulations were assigned to a revelatory paradigm, and modelling/problem solving to a conjectural paradigm; his other two paradigms, the instructional and the emancipatory are much as we have outlined them here. Other authors, such as Romiszowski (1986) have put forward alternative terminologies. In the rest of this chapter, however, we shall be more concerned with exploring the opportunities offered by computing developments to deliver various types of learning experience than in trying to classify them according to any particular schema.

3. COMPUTER MEDIATED LEARNING IN DISTANCE EDUCATION

As outlined in the previous section, CML can play a valuable role in any multiple-media educational package. For distance teaching institutions, however, 'new technology' has some very particular advantages and drawbacks. In Subsections 3.1 and 3.2 we offer a summary of the development of educational computing at the UKOU, which may serve to highlight some of the major issues.

3.1. Educational Computing at the UKOU

In the first decade of the UKOU's operation, during the 1970s, the institution's computing activity was almost exclusively based on mainframes. From the outset, the UKOU had operated a computer-based system for its student and tutor records, assignment handling and exam analysis. In this period its educational computing provision depended upon three mainframes situated in three fairly widely separated parts of the country. Students accessed these central computers through terminals situated in over 200 study centres, using the public telephone network.

Some of the problems with this type of provision arose from limitations of the hardware: initially the terminals were simple teletypes. Other difficulties were associated with logistics and cost: study centres were only open at certain times and in any case were not accessible to every student, and the telephone links were expensive. As more versatile – but more costly – hardware, such as high resolution graphics terminals, were introduced, it was clear that these could not be multiplied in a large number of sites. By the early 1980s, although most services were still based on the mainframes, microcomputers (nowadays more commonly called personal computers – PCs) were beginning to have a significant impact on teaching.

The most immediately attractive features of PCs were their graphics facilities, which were soon available in colour; graphics editors allowed screen images readily to be incorporated into interactive teaching/simulation software. In those days, however, fully interactive programs required state-of-the-art machines, and these were not cheap. It was therefore quite impossible for them to be permanently available in study centres throughout the UK. The only way to make them accessible to large numbers of students was to take them to residential Summer Schools – and that restricted their use to those courses for which there was a Summer School component. The additional difficulties of integrating CML into already crowded Summer School schedules, and the cost implications, meant that only a relatively small proportion of courses actually incorporated CML in their initial design. This gave each CML scheme something of the status of a pioneering project, with relatively little cross-fertilization between courses, although there were some useful developments in the creation of authoring languages and tools for the generation of CAL materials.

By the mid-1980s, the PC-based programs were being on the whole well received, and sales of copies on disc indicated that a surprisingly large minority of students actually had access to a personal computer at home or at work. The introductory computing courses, which had adopted the strategy of sending out small computers as returnable home kit items, were several times over-subscribed. The prices of personal computers were falling rapidly, and a wider variety of machines was becoming available. It was clear that the time had come for the University to develop a home computing policy, which would at last bring computing facilities to remote learners wherever they were.

Initial discussion focused on the choice of a machine, but after the collapse during 1985 of two major manufacturers the University realized the dangers of commitment to a single manufacturer. The home computing specification became not one of hardware but one describing capability and software levels. It was also regarded as essential that the software and hardware chosen be adequate to cover the complete range of courses (from Foundation to third level and in several different faculties) that would operate under the policy.

Alongside these specific discussions, a much more general debate was raging: although there seems to be a strong academic case for home computing, there were fears about its effect on access and openness. Would the addition of a computer, especially to a Foundation course, deter some potential applicants from enrolling, whether from doubts about their ability to cope or for financial reasons? The second of these barriers was removed at least to some extent by the creation of a rental pool of machines, with the annual rental fee set at the limit previously agreed as the maximum permissible costs for set books on any one course.

The home computing programme, introduced in 1988 on just three courses, had by 1990 already expanded to cater for 13,000 students on seven courses. The specialist support of these students (via a telephone help line), the production and despatch of disks (180,000 of them in 1990) and the continual work required to maintain, develop and ultimately expand the programme in response to rapid changes in technology, all represent substantial investments of capital and staff resource. Is such a programme worth the cost? One report charting the evolution of educational computing at the UKOU during the 1980s concluded that it was:

> The implementation of this programme ... reflects the University's strength in making use of established technology in an educationally innovative manner. The University has influenced neither hardware nor software trends but has followed the industry standards ... in the belief that good ideas for educational applications of computers last much longer than the hardware and software platforms that are available to support them.
>
> (Butcher and Greenberg, 1991, p. 20)

The 'educational applications' will be the subject of most of the rest of this chapter.

3.2. Computer Communication for Remote Students

Following the establishment of a home computing policy, the next step in the UKOU was the addition of modems to the specification, allowing students access to various types of computer conferencing and electronic mail systems. First used as part of a second level course *An Introduction to Information Technology: Social and Technological Issues*, computer communication is gradually spreading to other parts of the undergraduate programme.

It is no exaggeration to say that the provision of on-line CML can revolutionize the nature of distance education. For the remote learner, it is, like the more traditional delivery systems, free from constraints of time and place. But it has an almost unique ability to put the otherwise isolated student into a group learning situation. Even a simple electronic mail facility can greatly enhance students' one-to-one communication with their tutor. E-mail has the advantage over the phone of being asynchronous, so messages can be sent and received at convenient times, and of allowing pauses for reflection; it scores over the postal services in speed of response. However, the real power of on-line educational systems lies in the ability of computer conferencing to support group communication.

Detailed evaluation of the CoSy conferencing system used on the previously mentioned *Introduction to Information Technology* course was carried out in 1988, its first year of presentation, and the findings reported by Mason and others. It was found that only about a quarter of the students contributed to the conferences, with just over half of them reading or scanning at least some of the messages. Although some exposure to conferencing was obviously relevant in the context of this particular course, it was not built into the structure of the course and was seen by many as a 'frill' added on to the main teaching media. To exploit fully the educational benefits of conferencing, it would have to be seen by students to play an essential role in the delivery and teaching of the course. It could also prove a life-line for some students, in supplementing or replacing telephone and face-to-face contact with tutors and other students. The potential of computer mediated communication (CMC) in this respect is brought to life by two comments made by students during the evaluation, and reported by Kaye:

> . . . the CoSy experience adds up to much more than the total of all tutorials that I have attended on four previous courses . . . [I] feel more involved and a part of things than I have done on other courses. CMC . . . is one giant step towards removing that feeling of being 'on your own' suffered by OU students, certainly by me.
>
> (Kaye, Mason and Harasim, 1989, p. 32)

> The sort of help available via CoSy has made me feel part of a University for the very first time.
>
> (Kaye, 1990, p. 18)

The second quotation is from a disabled, housebound student and shows how CMC can make available activities, such as participation in discussion, that students with disabilities can obtain in no other way. In the next subsection we look briefly at some of the other benefits of computers for students with particular difficulties.

3.3. Computers for Students with Special Needs

Computers can be especially valuable in offering forms of input and output devices tailored to learners with physical or sensory disabilities. To help with inputting text, modified (e.g. braille) keyboards can be provided or replaced by a concept keyboard. (A concept keyboard is a large touch-sensitive board with a matrix of boxes which can be customized with the use of overlays for users who cannot cope with ordinary keyboards. Each box can contain a picture, a word, a letter, or even a phrase, depending on the overlay in use and so the box can represent a concept, hence the name.) For blind students, a voice recognition device can be used, and for those with impaired speech a speech-recognizing device. Headpointers can be invaluable for students with motor control problems, and can be coupled to a lightpen, keyboard or a concept keyboard. Output devices are also vital. They can include braille display or displays with enlarged characters or spoken signals. For more information on these topics see, for example, Hawkridge, Vincent and Hales (1985). The motivating effect of CML for students with learning or emotional difficulties are well recorded (e.g. Weir, 1987).

3.4. The Cost, Role and Potential of CML in Distance Education

A comprehensive literature review of the potential of CMC in distance education, including design and teaching considerations, institutional issues, and an overview of implementations of such networks worldwide, has recently been compiled by Wells (1992). A detailed review of the cost structure of various types of computing provision within a distance education package is not appropriate here. (An analysis of on-line costs, based on the UKOU experience, has been carried out by Rumble, 1989c.) However, general issues to be considered include hardware, transmission, development and software costs.

Hardware: In the 1990s, many institutions may feel that in order to teach certain subjects (notably, but by no means exclusively, those directly related to computing or information technology), students have to be given access to a computer. The UKOU experience related in Subsection 3.1 shows that a start can be made by bringing the students to the computers (or terminals); but for *distance* education the way forward clearly lies in getting personal computers to the students by one or more of the following schemes:

* loaning the machines to students for the duration of their course (high capital cost),
* renting the machines to students (thus recovering at least a proportion of the institution's capital outlay),
* requiring students to buy their own machines, which would have to meet a particular specification.

The third of these options moves the personal computer into the 'standard consumer goods' class, alongside the radio, TV, pocket calculator, and more recently video-player, on which many distance education institutions rely for delivery of their packages. The personal computer has not quite reached this status yet; the question for debate is whether it will do so in the industrialized countries before the end of the century; even then, it has to be recognized that students could not be expected to buy very highly specified machines. The medium may be emancipatory, but acquisition of the necessary hardware may be a definite barrier to openness.

Transmission: If CMC is to be part of the course structure, central computer costs, networking costs and connection charges all enter the equation. Transmission costs are particularly important, since they are borne directly by the students. On UKOU CoSy conferences, 'one person's self-expression is another person's phone bill' has become a watchword.

Development: The generation of CML is expensive in terms of personnel: Scanlon *at al.* (1982b) have estimated that it takes three months of professional person time to produce one half hour of interactive computer mediated instruction.

Software: Specialist programming effort is clearly essential for tutorial or

Generation	First	Second	Third	Fourth
Medium	Print or correspondence	Telephone or teleconferencing	CML	Networked computers
Means of communication	Visual	Auditory	Audio-visual	Audio-visual
Means of delivery	Mail	Phone lines	Computer terminal or PC	PC with modem
Type of instruction	Individual (either 'broadcast' [one to many] or 'tutorial' [one to one])	Individual or group [one to one or one to many]	Individual	Individual and group [many to many]
Mode of interaction	Asynchronous	Synchronous	Asynchronous	Asynchronous

Figure 8.2. Generations of technological delivery systems in distance education (adapted from Lauzon and Moore, 1989). Each successive generation can include any of the elements of the previous generation.

simulation packages, but there is a lot to be said for showing students how to get the most out of generic applications packages of the type originally designed for industry or commerce, e.g. spreadsheets, word-processing, data handling and graphics packages. Having been designed for a large market they sell relatively cheaply and often have more time spent on their development than many specifically educational programs. Such applications packages are revolutionizing the way scientists and technologists work, and should therefore be introduced to students as part of their courses.

The phrase 'as part of their courses' probably contains the key to CML in the distance education context. It is in the integration of computer technologies with other media that the immediate future lies: this is undoubtedly the best way to widen computer literacy. Thereafter, it will be progressively easier to increase the emphasis on CMC.

A number of workers (e.g. Garrison, 1986; Nipper, 1989; Lauzon and Moore, 1989 among others) have sought to classify distance education systems into various 'generations' according to their mix of media, instructional methods and modes of delivery. A composite diagram representing some of their analysis is shown in Figure 8.2. 'First generation' distance education is basically correspondence teaching. 'Second and third generation' systems have developed multiple media approaches but their basic tenor has been to foster *independent* learning. Their emphasis is always on the package of learning materials, rather than on the provision of a learning environment to cater for the needs of individuals. What the 'fourth generation' can promote is the concept of learning as a social process, with interactive communication within the peer group as an essential element. Some would even claim that this group-based learning, with its possibilities for active discussion and 'coffee-housing' (albeit electronically), at last brings the true university experience to the distance education student! Others consider that we are witnessing the birth of a new domain – on-line education – quite distinct from either the mainstream or distance education sectors. Recent global developments, such as the Internet and the World Wide Web will almost certainly ensure the rapid growth of this domain and this is an issue to which we will return in Chapter 15.

4. CML FOR SCIENCE STUDENTS: DATA HANDLING, SIMULATION AND MODELLING.

How easy it is to believe the computer always knows what it is doing just because it can do some things very well.

(Self, 1987, p. 46)

Many of the educational uses of computers described so far have been tested out in the teaching of science. Three applications in particular are interesting: data handling, simulation and modelling.

4.1. Data handling

One thing that computers do well is calculate quickly. So for real experiments, the computer can be invaluable as a means of displaying and analysing experimental results, or in the control of equipment, or in the monitoring of parameters during and the storage of data after the experiment. Using a computer saves time and gives students the opportunity of analysing for themselves more complex experimental data than would otherwise be possible.

4.2. Simulations

In a simulation, a program which models some process or system is made available to the student in the hope that by studying the performance of the program he or she will gain insight into what is being modelled. So, the computer is not simply a 'number-cruncher' but takes the place of a laboratory or of a field trip. Students can be invited to simulate experiments which would otherwise be impossible, or at least very difficult, for them to carry out for real, due to expense, time constraints, safety requirements or ethical considerations. For example, computer animations of reactions are a particularly useful adjunct to written teaching material on chemical or nuclear processes, when such reactions could not possibly be carried out by students. Simulations are also useful in considering chemical processes impossible to track in the laboratory because they go too fast (e.g. the rate of decomposition of hydrogen peroxide with a catalyst), or biological processes which are ethically or practically difficult to set up repeatedly (e.g. examining the effect of a pollutant on pond life or tracking genetic changes in a population). Using simulations, students can be given much greater freedom in the planning of their experiments than would ever be possible with real experiments. It can be argued that improved understanding of how science is done, and how data can be analysed and interpreted, should follow.

An example of a simulation that allows students to 'experience' the otherwise innaccessible world of high energy physics is the TRACKS program, developed for a UKOU second level undergraduate course (Every and Scanlon, 1983). The TRACKS program simulates particle tracks in an idealized computer-controlled bubble chamber. Students using TRACKS are given a description of a fantasy game where they have the goal of discovering a certain type of particle, for use in the drive of a spaceship, in order to escape from an alien planet. Playing the role of the ship's physicist, they examine particle collisions like the one in Figure 8.3 and apply reasoning about curvature in magnetic fields, conservation of energy and conservation of momentum in order to identify the mass and charge of the candidate particles. Students have to learn how to control the bubble chamber and interpret bubble chamber tracks to identify whether any of three commercially available particles on the planet are suitable for use in the ship's drive. Deductions about the particles' identities are made by looking at the length of particle tracks in the bubble chamber, observing the collisions the particles make and using a magnetic field to make measurements of charge and mass. The simulation has two main instructional aims: to improve students' understanding of conservation of mass, energy and momentum, particularly in collisions, and to get the students to solve the strategic problems involved in the design and execution of an experiment.

There are both advantages and disadvantages in the use of simulations. Of course the advantages of simulations in relation to learning are apparent outside the CML context. The particular advantage of the computer is its power and flexibility as a tool for controlling simulations. However, in teaching scientists, care must be taken not to overuse its power, or replace too much practical work with simulation, otherwise the learner may make unwarranted assumptions about the ease with which real experiments can be done or the ease with which parameters can be controlled. Another type of problem lies in the reality or otherwise of the simulation. In a simulation the model has already been created by the programmer and there are usually few opportunities to change the basis of the model. The student should be aware that he or she is using a simplification of reality created by someone else. This awareness can be turned to advantage if the underlying model which is used in the construction of the simulation is inspected explicitly by students and its status explained. The educational outcomes of enabling students to construct their own models are the subject of much current research, and modelling exercises will be discussed further in the next subsection.

Course designers must be aware of the dangers in replacing all experimentation, or too much of it, with simulation. Jenkins (1987), for example, has warned against developing computer models which are subsequently used to test out hypotheses without other non-computer based experimentation, regarding such a procedure as unscientific. He bases his argument partly on the fact that the models used for simulations are simplified versions of reality. On the other hand, scientific models usually have simplifying assumptions underlying them and explicit analysis of the model can be turned into a useful teaching point. Bentley and Watts (1989) extend this counter-argument, pointing out that students can develop the skills of prediction, hypothesis testing and model revision much more easily from computer simulations than from real experiments. The mere fact of access to physical equipment does not make the experimentation and model building necessarily more realistic or more complex; doing experiments while controlling one or two variables and ignoring other possibilities can also be too simplistic in certain settings. Ogborn (1987) goes further and claims that such computer exercises represent

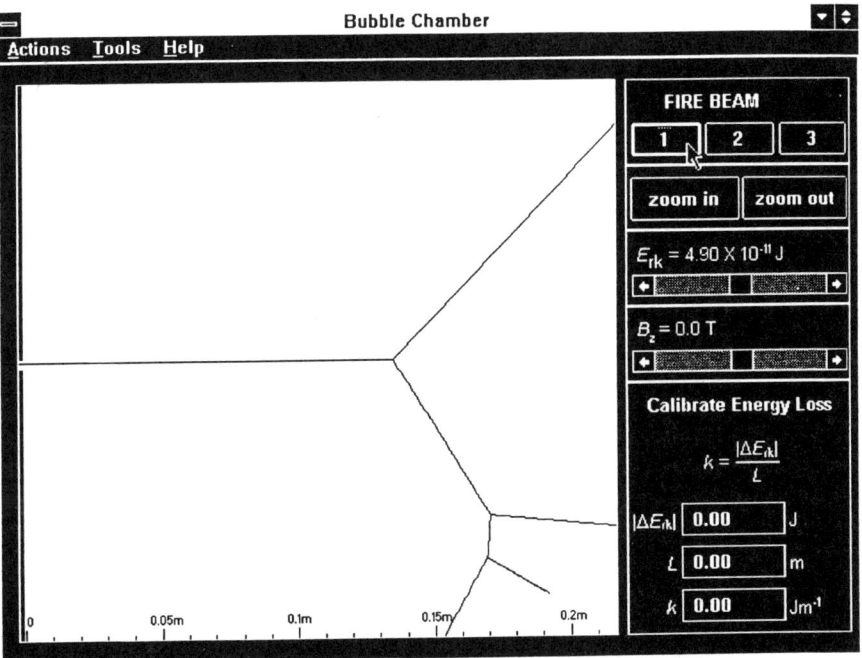

Figure 8.3. A screen image from TRACKS.

students' only first-hand experience of model building and theory creation. One can, of course, still take Jenkins' caveats to heart, while nevertheless recognizing that given the importance to science of producing explanatory models of how the world works, good simulations can help students to acquire the skills of scientific modelling. As a practical approach we would advocate the strategy adopted by the 'conceptual change in science' project (Twigger *et al.*, 1991), in which a curriculum for teaching basic mechanics has been designed to allow equal time for 'real' practical work and work with simulations. It is also important that, as far as possible within the limitations of their understanding, students should know what simplifying assumptions the simulation is based on and how questionable these are.

4.3. Modelling

The topic of model building has been raised already several times in this discussion of CML in science. Any software that allows students to develop their own (usually mathematical) models of a particular scientific phenomenon may be called a modelling package. In the present context, the term 'modelling' is best reserved for an activity in which the student looks into the structure of the model itself, rather than observes its behaviour as manifested in a simulation. It requires considerable skill to design environments in which students can create non-trivial models expressed as computer programs. It is therefore necessary for the student to be able to read, understand and change the program, and so student proficiency in some programming language may be an additional requirement.

An example of such a package is the Dynamic Modelling System (DMS). This flexible, general purpose modelling processor was developed by Ogborn and Wong (1984) at the University of London, and used in a number of English schools in connection with the revision of the Nuffield Advanced Physics course. It deals with models that compute the evolution of a system step by step. For example, step-by-step evolution is a good way of looking at the solution of differential equations, so Newtonian mechanics has been taught using DMS. The models are written by the student in BASIC. The screen editor allows initial values and lines of the model program to be easily altered. The program is flexible and can be used in a variety of ways, with pre-stored programs, modified ones or brand new ones, and when the models are run the output is in graphical form with a facility for tabulating values. The system also can be used at a variety of levels, from simply demonstrating pre-stored models, to learning about new situations by constructing models for them, to learning about classes of models and their relationships.

Modelling skills can also be developed using software not specifically designed for the purpose. The case of spreadsheets is particularly interesting. Originally developed as a business tool used for financial modelling, a spreadsheet controls an array of cells each of which have a numerical value which can be calculated from the values of other cells. The effect of a change in one independant variable on all other dependent variables is immediately displayed in tabular form as already noted in connection with Figure 8.1. Spreadsheets also provide a very convenient way of introducing science students to the kind of problems that are better solved by interactive numerical, rather than analytical, methods.

There is also a connection between spreadsheets and specially designed modelling tools. The Cellular Modelling System (Ogborn, 1987), designed for a younger age group than DMS, is like a spreadsheet in that the contents of an array of cells are calculated and displayed in a fixed sequence. This particular package has the added advantage of a graphical output.

There is much potential for exploring the uses of computers for modelling in science education. Although one may have reservations about the desirability of students having to learn even a simple programming language such as BASIC, first, with the availability of these sorts of tools, the onus is on science educators to make modelling with computers a really exciting, as well as an educationally productive, exercise for their students.

5. THE FUTURE

With CML two forces are at work at any one time. One is the pace of technological and economic change. New computer developments combine with rapidly dropping costs for ever more powerful equipment to create what Hawkridge calls 'creative gales of destruction' (1990). These developments may have a great impact in the educational world, but are not usually driven by it: in both hardware and software terms the agenda for change will be dictated largely by business users. The second force is the influence of educational research into new ways of using computers. Some of the issues being tackled emerged from our previous discussion. For example, the deficiencies we described in the feedback provided by tutorial programs can be addressed by using techniques developed by researchers on intelligent tutoring systems, or computer based instructional systems that display some degree of intelligence (O'Shea and Self, 1983; Wenger, 1987). The requirements necessary for improved adaptivity in tutoring are that the system is not *provided* directly with teaching materials, but *generates* those materials from its knowledge of the domain, teaching techniques and the history of the individual student. The intelligent tutor system thus can reason about its own effectiveness and adapt as necessary. It is difficult to find operational examples of such programs though one good working example is STEAMER (Williams, Hollins and Stevens, 1981), an instructional tool for training engineers who will operate large ships, which was in use at the Great Lakes Naval Training Center. It consists of an interactive inspectable model of the steam propulsion plant which students can experiment with, and includes

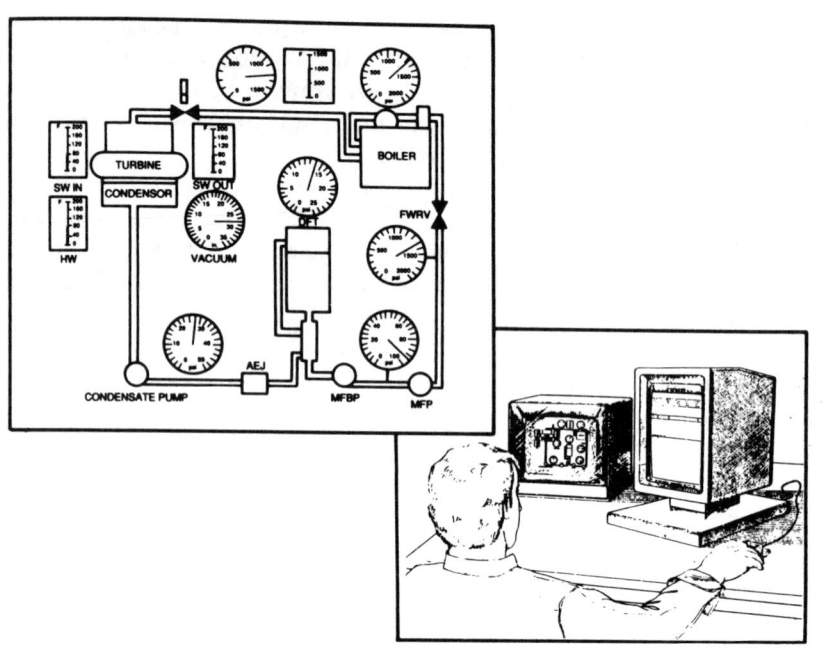

Figure 8.4. A screen image from STEAMER (Wenger, 1987, p. 81).

a minilab for experiments and a tutorial module that provides feedback during the execution of known procedures (see Figure 8.4). STEAMER is in fact an example of an interactive learning environment, a term currently more in vogue than tutorial or simulation. It encapsulates both types of program offering various degrees of learner control.

Many of the problems associated with research into tutoring systems have been overtaken by events which have changed the technology of the delivery devices. For example the dialogue between student and machine was the subject of study from the 1970s onwards using artificial intelligence natural language processing systems. However, the graphical interfaces which have been introduced subsequently (e.g. WIMP, Window, Icon, Mouse, Pulldown menu systems) are easy to use, and largely circumvent many problems of natural language understanding. These and other graphical interfaces, arising out of research conducted at the Xerox Palo Alto Research Center, have changed the potential of student simulation work. For example, the Alternate Reality Kit (ARK), a graphical animated environment for creating interactive simulations constructed there (Smith, 1986) incorporates some unique features chosen to make its simulations more powerful and convincing. The package is based on a physical metaphor: all objects have a screen image, a velocity and can experience forces. A microworld consisting entirely of these objects can be constructed to facilitate the understanding of basic physical laws. The consequences of breaking the laws can be seen by manipulating the objects directly on the screen. The proven benefit of directly manipulating concrete objects like Dienes blocks is well known in mathematics. Research workers in science education currently hypothesize that in constructed microworlds, the direct manipulation of the objects and processes represented will help learners to internalize the concepts appropriately. They also hypothesize (O'Shea, O'Malley and Scanlon, 1990) that counterfactual reasoning and learning by conflict resolution are useful teaching techniques for work with simulations. Counterfactual reasoning involves imagining that the laws of maths or physics were otherwise and running simulations to see the consequences. Conflict resolution techniques confront the fact that students often hold several conflicting models to account for their understanding of a domain, including persistently erroneous models. It is necessary for learners to face the consequences of conflict between the models they hold.

An example of an environment based on ARK and incorporating such principles as counterfactual reasoning is Friction World (Spensley *et al.*, 1990) shown in Figure 8.5. In seven different alternate realities, simulated worlds named after the planets, experiments in sliding friction can be carried out. One of these realities behaves like the real world. Each of the others is based on a misconception about friction, and behaves differently; they were named after the planets to emphasize the distance between the simulations and the real world. The 'experiment' that students carry out involves dropping a block down a chute so that it is propelled across one of three surfaces with different coefficients of friction. (The material that makes up the surface is altered by students dropping either sand, syrup or ice from hoppers suspended above the floor.) As the block falls, the various meters shown on the screen record the height from which it was dropped, the time it took to fall, the velocity with which it fell down the chute and the horizontal distance travelled. The interface objects are moved using a mouse. This system has been used successfully with students at a UKOU mathematical modelling Summer School and suggests a wide range of future experiments. Hartley, Mallen and Byard (1991) have described the use of VARILAB, a qualitative modelling tool which allows school pupils to express their own models (whether correct or incorrect) of the cause and effect relationships that hold within a given mechanical system.

Most of the discussion in this chapter has been about working systems already in use in educational settings, where sufficient experience of them has been gained to allow some meaningful evaluation of their learning outcomes. This implies necessarily that some of the programs we describe might even seem dated. However, the speed with which educational computer related innovations become sufficiently developed to be usable by students is not that great, partly because the technology of delivery devices in widespread use changes only slowly. However one recent technological development that has already begun to make significant impact on educational uses of computers is the increasing availability of multimedia and we deal with the potential this development has for distance education in science in the next chapter.

One further point worth mentioning here is the growing interest in co-operative rather than competitive learning and the role that computers can play in its development. Systems designed to encourage shared work, such as SharedARK (Smith, 1993) may even serve as a prototype for future distance education. In SharedARK a single user site consists of a computer workstation, an audio headset, a video monitor and a videocamera. Each site is physically isolated from the others but connected to them via a computer-switchable audio–video link and high performance computer networks. The system is intended to support conventional distance learning activities, but also to provide the possibility of interaction between other students and tutors over video links, or even of distributed interactive simulation laboratories. Currently experiments with such systems are focused on identifying the features of successful collaboration via computer (Smith *et al.*, 1991). The associated technology is currently too expensive for widespread use, but the experiments give some flavour of a potentially exciting educational development.

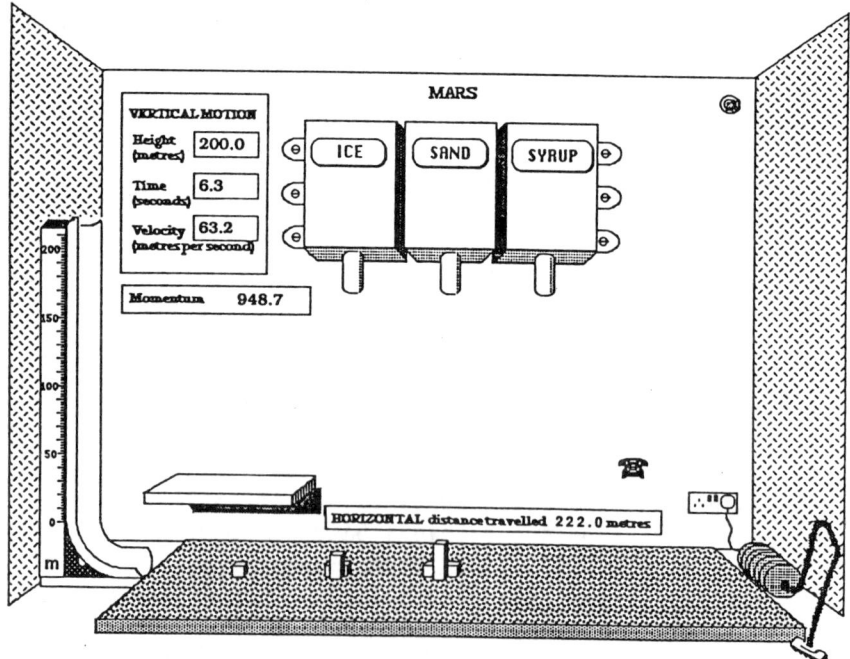

Figure 8.5. A screen image from FRICTION WORLD. The task involves dropping bricks down a chute, across one of three surfaces with different coefficients of friction. (Spensley *et al.*, 1990, p. 20).

9

MULTIMEDIA

We need a rich approach to the message and the medium. The message is not just 'content', because a deeper analysis of the knowledge is required. The medium is not just a technical format, such as 'video' or 'CD-Rom'. It is the whole presentational style, the user interface, the accessibility, the interactivity.

(Eisenstadt, 1995, p. vi)

1. INTRODUCTION

The dividing line between the kind of CML discussed in the previous chapter and computer-based 'multimedia' technologies is that the latter offer the ability to incorporate real-world imagery into the package. This adds a new layer to simulations, with the inclusion of still or moving pictures of actual events and places. Multimedia technology offers the (thoeretically) seamless integration, with a graphic user interface, of screen text, still graphics and photographs, animations, sound and motion video, all at the click of a mouse. The benefits of TV and video for science education discussed in Chapter 7 are all present in multimedia packages, but with the huge additional advantage of interactivity. Students can participate in vicarious experiments, not just by taking data from the screen, but by actively choosing one experimental strategy over another – and can do so without any risk to themselves, should their choice be a poor one. They can be taken on virtual field trips to otherwise inaccessible locations and actually select for themselves the parts of the environment they want to concentrate on at any particular time, and the route by which they explore the landscape.

The degree of control given to the learner in using multimedia packages can vary from minimal to total. The three-tier classification scheme proposed by Romiszowski (1984, 1986) in relation to instructional design in general is particularly appropriate for multimedia. Under this scheme, the two ends of the spectrum are represented by 'autocratic' and 'democratic' control modes, with a 'cybernetic' approach occupying the middle ground. The *autocratic* mode is totally prescriptive and didactic, with the routes through the material set by the instructor and forced on the student by the software. These paths may be branched, but the learner is nevertheless directed to follow a particular sequence in using the package. As discussed in Chapter 8, this kind of pre-programmed instruction has its place in drill-and-practice routines, but does not sit well with constructivist views of learning. The *democratic* control mode allows the students to use the material in whatever way they wish, to fit their own learning strategies. The multimedia package becomes an enormous random-access databank-cum-encyclopedia; many of the early releases of CD-ROM for the 'home' education market, such as Microsoft ENCARTA, were of this type. The balance of control in most multimedia systems designed specifically for instruction lies somewhere in between these two extremes; *cybernetic* systems also offer individualized on-line advice, according to an analysis of the student's choices and performance while working through the package, and what is therefore perceived to be their ability, skills and knowledge base, and personal learning strategy.

2. VIDEODISCS: FIRST GENERATION MULTIMEDIA

The first multimedia programs to be developed in the educational field were based on computer control of video systems, intially videotape and then videodisc. The great attraction of these programs lay in their ability to combine the high quality audiovisual characteristics of the video element with the graphics display and versatile interfacing capabilities of a computer; the videodisc technology further scored over videotape by virtue of its high storage capacity, ease of display of still frames, and random-access capability. For the first time, all the elements of interactive CML discussed in the previous chapter could be combined with audiovisual components that could realistically simulate a wide range of environments for the learner to explore.

In the UK, the development of educational videodiscs in the early 1980s was heavily funded by the government, through the Interactive Video in Schools (IVIS) programme, and through Department of Trade and Industry grants to selected projects in higher education. Many of these grants went towards the production of videodiscs aimed at the training of doctors and other health professionals, but the UKOU was a recipient of funding for a videodisc produced as part of a second level physics course on the properties of matter. This particular disc – the water videodisc – was used to give students some practice in problem solving. The videodisc technology has, as Laurillard (1987) has pointed out, huge strengths as a medium around which to structure problem solving. This can be a particularly difficult set of multi-layered skills for remote learners to practise because of the constraints on their access to information and materials. Videodisc technology combines an interactive, non-linear CML with a random-access disc that can provide high quality visual presentations. It is therefore particularly suited to helping learners explore the rather open-ended type of real-world problem, which is often best approached in a non-linear way and involves pulling together many different types of information. The process of solution involves a number of activities relating to different steps, which may be carried out in a fairly random order – collecting data from reference books or by carrying out an experiment, 'reading round' the subject, drawing diagrams, deriving equations and performing calculations. In the real world, one can move from context to context in these various steps, while keeping one's place in each. In a multimedia environment, the computer maintains the various contexts, and saves them as the user moves on to a new context. It is thus possible to roam freely through the resources of a videodisc, saving the state of motion of video sequences that have already been accessed, data that has been retrieved and any previous alpha-numeric inputs.

In the water videodisc, the problems that the students are invited to explore all relate to the physical properties of ice, water and steam: for instance, is it a feasible commercial proposition to tow an iceberg from the Antarctic to Australia in order to provide fresh water for agricultural irrigation? (The other problems posed on this particular disc have been described by Bolton, Every and Ross (1990)). In order to tackle the iceberg problem, students have to think carefully about how they might model the situation, and about what assumptions and simplifications might enable them to make headway. They also have to bring into play many important physical concepts and processes. For example, one of the major steps will involve an estimation of how much mass is lost from the iceberg during its passage; this in turn requires sub-steps that quantify the heat absorbed by the iceberg from both the sea and the incident solar radiation. Students therefore have to search the 'library' information files or perform experiments in the virtual 'laboratory' to obtain the necessary input data (e.g. sea and air temperatures, radiation levels, reflectivity of ice, specific and latent heats, etc.). There are more than enough resources on the disc to solve the problem, and the students may access these more or less at will. However, in order to make the disc reasonably free-standing for the remote learner, there is some guidance provided in the form of a narration on video, which helps the user to identify the main steps on the route to a solution, and supplies some encouragement to continue. The software checks all intermediate inputs, and provides appropriate tutorial comment or context-specific help. Evaluation of the water videodisc, reported by Bolton *et al.*, showed that students were enthusiastic about both the medium and the package. The varied multimedia presentation and the requirement for continual active response kept them involved for long periods, and they derived considerable satisfaction from solving complex real-world problems 'on their own'.

Videodiscs have also been used successfully to provide vicarious experience of practical work, based on simulations of laboratory and field environments; these will be described in Chapter 10, and we will outline other multimedia projects with similar aims in Section 3.

Videodiscs can be viewed in some ways as a pioneering technology – providing the initial means by which different types of environment could be realistically simulated within a single interactive learning package. Videodiscs spread gradually into the non-formal educational sector – museums, exploratoria and science centres – and during the late 1980s it seemed that they might become big business in the American high school market. However, their distribution was always held back by the high cost, and consequent shortage, of the playback hardware, coupled to the exceptionally high demands of the medium in terms of development resource. By the end of the decade, it was clear that this situation had become a

> vicious circle of there being too few discs available to justify the cost of expensive playback machines, and yet little incentive to develop discs which . . . [could not] be disseminated because of shortage of players.
>
> (Ross, 1991, pp. 102–3)

Research into learning and instructional design theories as they might be applied to this new medium was of course similarly hampered. Such problems are only now being gradually overcome by the spread of multimedia packages based on CD-ROM, a technology that can deliver similar or even enhanced interactivity but for which the hardware is becoming increasingly widely available.

3. CD-ROM: SECOND GENERATION MULTIMEDIA

> Experience in grasping or recreating the structure of a knowledge base is an important part of a learner's development . . . Too often we hand the learner an itinerary rather than a map of the terrain and a survival kit. This partly stems from the logistical problems posed by handling authentic source material and raw images. To solve them, we not only need powerful and friendly software tools, but also very large and robust multimedia storage. These needs are respectively met by hypertext and compact disc media.
>
> (Megarry, 1989, p. 50)

The 'hyper' concept was first discussed in the 1960s (see, for example, Nelson, 1967), long before computers had the capacity to handle text. The operating principle of hypertext is quite simple – it is essentially a cross-referencing system. Any word in the text can be selected as a trigger to link to other text. The essence of multimedia is that these links need not be just from or to text, but can be effected to and from anything stored on, or interfaced to, the computer – graphics, video, sound, other programs. At the moment, CD-ROM provides by far the most convenient method of multimedia storage.

While in many respects laser videodiscs and CD-ROM offer similar pedagogic advantages, CDs score over videodiscs in a number of ways. The CD is an entirely digital technology, in contrast to a videodisc, on which images and sound are stored in analogue format. Digital coding facilitates data integration, interactivity with the video information, and the reconciliation of the different television standards operating in different parts of the world. Audio-CDs are a mass-market consumer product, and consequently disc pressing has low unit costs. The acceptability of audio-CDs, the small size of CD-ROM drives and the ease with which they can be slotted into PCs, have paved the way for the CD-ROM revolution to be forged in the home and leisure market.

One example of a CD-ROM package currently being developed by the Biodiversity Consortium (a group of British universities, including the UKOU and BBC OU Production Centre as a key production site) is a 'desert field trip', and a description of this package may give a flavour of its educational potential. The students are transported to the Sonoran desert via a screen-based workspace. They are presented with a variety of desert panoramas, showing both daytime and night scenes, all filled with animals, plants and other features for them to locate and identify. They can select the equipment they want to take with them, and can navigate the environment at their own pace and in their own choice of direction. Free-ranging exploration of the virtual realities of multimedia is greatly facilitated by the network of links that can be set up between 'hotspots'. The user can click a mouse pointer on an animal or plant somewhere in the desert landscape for example, and be given a close up picture of the organism, a motion video sequence showing how it moves, flies or disperses its seed, a textual description, and a series of cross-references to its place in the food chain. Clicking on a word in a text display takes the student on to further definitions, screen images or connections. Learning becomes a process of discovery, in which material can be browsed rather than read sequentially. An electronic notebook enables students to keep a record of their observations and conclusions. A 'satellite' icon allows them 'remote' access to relevant data. They can also call upon the services of an expert 'tutor', so as to receive context-specific help, pointers to important features they have missed or questions to set them thinking along the right lines. What may begin at first as a 'natural history tour' soon develops into a much more detailed scientific analysis (for which tools such as calculator, glossary and graphical facilities are available on screen). Objectives are listed on a pull-down menu, to make it clear to students what they are expected to gain from the field trip – not just the ability to recognize desert species, but also, for example, an understanding of the adaptations that fit those organisms for life in the desert environment.

Pilot projects are underway in chemistry for problem-based CD-ROM 'laboratories', in which students will be set a practical problem, and with an element of free choice will be guided towards finding a workable strategy, and the right equipment and techniques, to solve it. One very important strand in this package will be the requirement to use the database facilities to discover and evaluate the safety implications of any proposed strategy. The high storage capacity of CD-ROMs also allows a pattern of resource-based learning very different to any that has been feasible for remote learners in the past. 'Libraries' of source documents, research papers and abstracts can now easily be provided, in which students can carry out systematic literature reviews or simply 'browse'.

As already mentioned in Chapter 8 in connection with earlier generations of CML, CD-ROM technology can be adapted to be of considerable assistance to students with special needs, providing for instance large print displays for visually impaired learners. Such developments are not infrequently found to have unexpected spin-offs. One example is the 'virtual microscope' (Robinson, 1994a). As originally conceived, the virtual microscope was designed to allow disabled geology students to have the experience of using a polarizing microscope. Basically it consists of a CD-ROM containing thousands of images or rock thin sections seen from different angles, at different magnifications and with different degrees of polarization. However, because the virtual microscope offers such clear images and the opportunity to see polarized and unpolarized views side-by-side (a feature not available on real microscopes), it gives new insights that have been found to benefit all students of Earth science. It also allows students to concentrate on geological features and their interpretation, without being hampered by poor microscope technique (e.g. wrong level of illumination, non-orthogonal polars, etc.). It is not a substitute for real microscope work, but a powerful learning tool, which enables students to make best use of limited microscope time.

4. PROBLEMS AND POSSIBILITIES

The requirements for team work in putting together a learning package, already discussed in Chapter 4 in relation to mixed-media courses, applies *a fortiori* to multimedia packages. Learners used to the visual sophistication and narrative strands of TV and video will not take kindly either to crude graphics produced by someone without professional expertise in visual design simply using the toolset of an authoring package, or to a video presentation that does not flow.

> We should not forget that the word multimedia implies multi-disciplinary approaches; a one-dimensional approach, whether by a techie (and we all share, I would have thought, the perceptions that no [educational] developments should be technologically-led), or an academic . . . is bound to result in a one-dimensional product.
>
> (Fowler, 1995, p. x)

As with teams producing more conventional educational material, there is a balance to be struck here between the desire of academics to retain 'control' of the content and teaching style of the material, and the desire of the media professionals to generate an attractive 'product'. We will return to this general theme in Chapter 12.

Although the technical difficulties of multimedia production are being continually reduced by the development of powerful authoring tools, and although many of the costs associated with production (e.g. hardware, prototyping and CD-ROM publishing) have fallen considerably in recent years, multimedia packages are nevertheless very expensive to generate, largely due to their ravenous appetite for audiovisual material. If this material is specially originated, the involvement of professional media specialists and the time required result in rapidly escalating costs; if it is imported, the copyright issues can be complicated and expensive to agree. The 'choreography' of a multimedia package is also a complex matter, very consuming of academics' and producers' time. 'It is not difficult, therefore, to see why we have many glitzy examples of the exciting *potential* of multimedia but few serious packages that meet higher education learning needs' (Darby, 1992, p. 2).

Another factor hampering the spread of multimedia packages into the university sector is the fact that the 'not invented here' syndrome is still strong in many institutions of higher education and, as noted by Laurillard, Swift and Darby (1992), Darby (1995) and others, particularly so in respect of CML programs. Academics who would be willing to put a textbook on a reading list, because they could give its content their own individual slant in lectures, seminars, or supplementary course material, are far less willing to include any kind of computer package from other sources. This resistance probably arises from the fact that most of the teaching structure and style are already built into computer-based material, and the possibilities of mismatches in philosophy are legion.

The solution for both these problems currently being adopted in the UK is government funding of the first stage of the Teaching and Learning Technology Programme (TLTP); as mentioned in the previous section, this aims to use consortia of many institutions to produce multimedia materials that will then be distributed widely throughout higher education. However, there are those who argue that this way of generating courseware is only a short-term solution to a far longer-term problem. For multimedia programmes truly to penetrate the higher education environment, their development must be self-sustaining. Whether that can be achieved will depend on volume, as well as pedagogical perceptions. Development of new packages will only be viable without subsidy if enough students and institutions make regular use of them.

One way forward may be to make introductory and service courses in higher education institutions, for which there is already fairly broad agreement on content, even more uniform. In his 1991 Rede Lecture, Sir Peter Swinnerton-Dyer, former chief executive of the (UK) Universities Funding Council, advocated this path:

> We need to consider very seriously whether we can harness modern technology to provide economies in the process of teaching and learning. Nothing would be lost if the first course in abstract algebra or thermo-dynamics were everywhere the same.
>
> (Swinnerton-Dyer, quoted in CTISS report, 1992, p. 11)

An alternative scenario is that it will become necessary to design much more flexible materials that can serve a variety of markets on an international basis and be repurposed at each point of use. One possible approach is to bypass entirely the idea of matching individually crafted multimedia packages to individually created courses, and to use multimedia materials purely as a learning resource (in similar mode to textbooks, dictionaries, encyclopedias and journals), harnessing the power of communications technology to make these materials globally available on the Internet:

> if interactive multimedia as a teaching tool is to fulfil its potential, it must be wrenched from its stand-alone position and moved firmly into the domain of global telematics. The possibilities are then endless for all aspects of distance learning and the dissemination of information.
>
> (Jacobs, 1992, p. 15)

This view represents one of the possible futures we will explore in Chapter 15.

10

PRACTICAL WORK

A fool . . . is a man who never tried an experiment in his life.
(Erasmus Darwin, 1731–1802)

1. INTRODUCTION

Lord Perry, the first Vice-Chancellor of the UKOU, recounts that when, during the setting up phase, he spoke in public about plans for the new institution, one of the questions he was constantly faced with was: 'How are you going to teach science subjects, for which a significant proportion of the course must be founded on laboratory work?' Sceptics pointed to the fact that in mainstream undergraduate science courses, a large amount of time is devoted to practical work carried out in the laboratory (or, for some subjects, in the field). Was it not therefore the case, asked the sceptics, that a distance education system, by freeing students from compulsory attendance at set times or places, must be incapable of providing its science students with the necessary practical experience?

Perry's response, rooted in the university's multiple media instructional strategies, was that the practical element could be covered in a number of different ways, of which at that time the most important were 'home experiment kits', demonstrations on television and residential schools. Twenty-five years on, these are still crucial components of many UKOU science courses, although a fourth strand, involving various computer mediated technologies has now been added. The different and overlapping roles played by these various media in teaching and learning scientific 'practical skills' will be explored in later sections of this chapter.

However, before looking in detail at the devices and delivery systems that may be used to offer the remote learner an approximate equivalent or indeed an alternative to laboratory experience, it is appropriate to examine the aims of science educators in providing such experience for their students. From current perspectives, it is interesting to note that the fledgling Open University documented by Perry did not challenge at institutional level the assumption that a substantial laboratory component must be an essential part of an undergraduate programme in science. The acceptance of the then standard view on this issue was certainly closely tied to the battle for recognition and academic respectability: the concept of a distance teaching university was still sufficiently revolutionary for it to have been unwise also to introduce radically different curriculum criteria. Nevertheless, Perry records that many academics did, even then, privately question the need for such heavy emphasis on practical work. These kinds of doubts are once again very much to the fore, with the debate about the aims and effectiveness of laboratory instruction becoming ever more lively. We will therefore begin this chapter by exploring some of the principal issues in this debate.

2. THE PRACTICAL ELEMENT IN TEACHING AND LEARNING SCIENCE

2.1. Why do practical work?

In the mainstream sector, a large amount of time and teaching resources for a subject such as physics, chemistry, or biochemistry – typically between a quarter and a half of the contact hours – are devoted to laboratory-based work. So what are the purposes of this emphasis on laboratory experience? A number of classifications for the aims of practical work have been presented in the literature (see for example Henry (1975), Black and Ogborn (1979), and Woolnough and Allsop (1985), among others). An amalgam of these approaches suggests the following clusters of aims:

1. Training in techniques. This cluster relates to the acquisition of

    (i)   skills in the psychomotor domain (i.e. manipulative skills), which allow safe and appropriate handling of particular types of standard apparatus;
    (ii) skills of observation, measurement and estimation, during the acquisition of which students should develop 'good habits', such as accuracy and honesty in recording results, a systematic approach and so on;
    (iii) skills of data handling, interpretation and evaluation.

2. Learning the basic ideas of the subject. This cluster relates not so much to skills development as to the carrying out of experiments that introduce, illustrate or reinforce concepts and theories taught in other parts of the course. Sadly, such experiments often require students to do little more than reproduce a standard result, and are all too frequently designed according to a 'recipe book' model. A less sterile, more holistic approach (Woolnough, 1991) emphasizes the importance of giving students a 'feel for phenomena', thus allowing them to build up personal and practical experience in their subject, and to develop some appreciation of the complexity of the real world.

3. Learning how to do research. This cluster invokes more abstract sets of skills, involving an appreciation of scientific methodology and an ability to conduct an experimental inquiry. It is probably necessary to achieve a certain competence in many of the areas listed under the first two clusters before these higher order objectives related to science as a problem-solving activity can be adequately pursued.

4. Motivating through interest. This cluster contains aims in the affective domain, based on the premise that practical work can be important in influencing attitude development, with enjoyment of practical work leading to greater confidence and increased motivation. Arguably, the younger the students, the more crucial are the aims in this cluster, but they do have a particular importance for distance education students of any age, who are necessarily cut off from some of the potentially motivating contacts with practising research scientists enjoyed by students in the conventional university sector.

2.2. Training Scientists: The Paradox of Practical Work

The acquisition of basic skills of the 'scientific craft' – the first cluster listed above – were researched with an early generation of UKOU students (Aspden, 1973). A group of Foundation level students was divided into those who did and those who did not have prior laboratory experience. Comparison of the performance of the two groups during a residential school practical session showed that the experienced students exhibited no more competence than the inexperienced ones in most of the core skills – designing the best set-up with given equipment, collecting and analysing data, drawing graphs, etc. The only advantage the experienced group had over the non-experienced was a greater confidence in handling laboratory apparatus. Here indeed is a paradox: received wisdom has it that one cannot become a practical scientist without serving one's apprenticeship in a laboratory, and yet it seems that many traditional laboratory courses fail to equip students with even the most basic practical skills.

This issue goes right to the heart of the current debate about the purpose and value of laboratory instruction. Practical courses, especially those that are laboratory based, are expensive – not only financially but also in terms of staff and student time. The commitment of so much resource to laboratory instruction is only justified if its educational benefits are clear, and the resource is managed to maximize those benefits. Several authors have recently pressed the case for a re-evaluation of the aims and objectives of practical work, across the science curriculum from secondary (Woolnough, 1991) to tertiary (Toothacker, 1983; Stokes and Stafford, 1986), and including the distance education sector (Kirschner and Meester, 1988).

One of the problems with the conventional laboratory teaching programme is the sheer breadth of its aims. In practice, many of the skills associated with experimental investigation are never explicitly taught; students are somehow expected to pick them up by 'osmosis', simply through the experience of working in a laboratory environment. As shown by the Aspden study and others (see, for example, the references cited by Toothacker (1983)), the only skills that are demonstrably acquired solely through laboratory experience, are the specialized manipulative sort: for example, the ability to use a microscope is the kind of skill that can only be attained through practice. There are also other, less easily categorizable, competencies that are heavily dependent on experiential learning: a feel for the magnitude of physical quantities, the quickest way to track down the fault in malfunctioning equipment, or the differentiation of 'rogue data' from unexpected results worthy of further investigation – to name but a few. Other skills, even though they are normally classified as practically based, may in fact be better developed through separate, specially designed exercises. For example, it is not essential for students to have taken measurements themselves in order for them to learn and practise skills of data handling or hypothesis testing. Indeed there may be distinct advantages, especially in the initial stages, in developing such skills away from what Woolnough and Allsop describe (p. 44) as 'the distracting clutter and inevitable inaccuracy of student practical work'.

2.3. Conveying a Practical Flavour in Distance Education Courses

The growing realization that a laboratory is not necessarily the best place to pursue each and every one of the practical objectives associated with a scientific training is good news for the distance education sector. Restrictions on laboratory access and time do not represent major obstacles when alternative teaching and learning strategies are available to assist science students in developing their skills. The distance education sector may even have an advantage over the mainstream in developing such stategies, in that institutions in which whole course packages are assembled from mix-and-match sets of components offer particular scope for the development of innovative instructional techniques. In addition to text-based materials, the practical element can be fostered by:

- home experiment kits;
- radio or audio-cassettes (the latter usually in conjunction with text, diagrams, maps or specimens, as an audiovisual sequence);
- televisual media (broadcast TV or video-tape distributed in cassette form);
- computer mediated systems (including simulations and multimedia technologies such as interactive videodiscs or CD-ROM);
- laboratory classes (either in the form of local day schools, or as part of residential schools).

It is not necessary, or indeed even desirable, for a single course to include all of these types of courseware. Each medium has its particular strength as a teaching or learning tool, which can be imaginatively exploited to help students achieve practically oriented objectives; the skill of the academic or educational technologist lies in choosing the one most suited to putting across a particular message. In the next few sections we will illustrate some of the ways in which various types of courseware may be used to give either hands-on or vicarious practical experience to remote students.

3. HOME EXPERIMENT KITS

3.1. The nature of home experiment kits

As noted in the introduction to this chapter, home experiment kits (HEK) were part of many of the UKOU's science courses from the inception of the institution. This is a set of items sent to each student on a course, for his or her exclusive use during that particular course. It contains all the specialized equipment and consumables required to complete the practical exercises within the course, although students may need to provide items of the type normally found in the average household (e.g. scissors, matches, tape measure, sugar) or of the sort easily available without recourse to specialist shops (e.g. distilled water). Kits represent a substantial capital investment for the institution, and for this reason most of them have to be returned by students at the end of the teaching year; they are then refurbished and sent out to a new set of students the following year.

Kits may be large or small, complex or simple. Some kits designed at the UKOU have consisted of a single core component (such as computer hardware), around which all other parts of the course are designed. Others have contained a broad range of items, supporting a variety of experiments. A few have even incorporated living organisms: a *Schizophyllum* (fungus) culture and live *Drosophila* (fruit flies) have both been included in HEKs for UKOU genetics courses. Chemicals, glassware, microscopes, rock samples, electronic components and balances are all common ingredients of UKOU science kits. The largest kit ever produced in the UKOU was the one assembled in 1971 for the first version of the Science Foundation course: this contained 272 separate items, and provided material for 30 experiments covering physics, chemistry, biology and Earth science. Nowadays, however, there is an increasing trend towards miniaturization (Waterman and Thompson, 1995), and disposable items that save both the inconvenience to students of having to return kits and the cost to the institution of having to refurbish them annually. A kit recently produced at the UKOU for a second level course on *Our Chemical Environment* contained over 50 chemical samples and items of equipment, allowing students to undertake over 30 very varied experiments, yet overall was smaller than a shoe-box and cost less than £5 at 1995 prices.

While in general HEK equipment needs to be as simple and user-friendly as possible, UKOU experience has shown that kits containing one large, expensive piece of equipment are often particularly successful, especially if this item is used repeatedly and/or is perceived to aid understanding of large sections of the course. For example, kits containing a microscope and thin sections are rated very highly by geology students. In large measure, this reflects the importance, within the subject context, of the information available from thin sections. But the 'street cred' factor comes into the equation too: such a kit has an effect in establishing the part-time, remote learner as a 'real' geology student – suitably equipped with microscope. A related phenomenon is the 'glamorous item' such as the laser and 'make your own hologram' package that feature in the kit for the second level UKOU *Images and Information* course. Although this HEK has a very serious purpose, in that completion of the experiments is essential for an understanding of many of the most important concepts in the course, a significant number of students are attracted to the course principally because they will get 'their own' laser, albeit just for a year, and the equipment to manufacture a hologram themselves.

At the other end of the spectrum, it is also worth noting that in designing experiments to be done at home, it is not always necessary, nor indeed even desirable, to think in terms of elaborate equipment. As Cornish, Jenkins and Atkinson (1981) has pointed out, it can actually be beneficial for a proportion of experiments in some courses to be based around household materials if this emphasizes in learners' minds the relevance to everyday life of the material they are studying.

3.2. Experiments for remote students

Lord Perry reported that when he first described the HEK concept to audiences of academics, they often tended to dismiss it as little more than a kind of child's chemistry set, capable of sustaining only a small number of rather stereotyped experiments. The short-sightedness of such views has since been amply demonstrated by the sheer variety of home experiments that have been incorporated into UKOU science courses. Examples of kit items for different subjects and some outlines of HEK-based projects may serve to illustrate this.

Earth science kits containing thin sections and a petrological microscope have already been mentioned. Other geology courses have included extensive practical work using geological maps, rock and mineral specimens, and instruments such as compass clinometers, all supplied in kits. A third level course about evolution involves a HEK morphology exercise on the evolution of a lineage of Brachiopods from the Jurassic, in which students make measurements (using calipers) of both real and replica fossils.

Chemistry students often receive a molecular model kit, which they find a very useful aid to spatial visualization of chemical structures. Otherwise, chemistry kits usually consist of the standard kinds of equipment – glassware, clamps and stands, tripods, burners (fuelled by camping gas cartridges), and chemicals. The inclusion of a balance allows quantitative determinations: for instance, Foundation level students can carry out a stoichiometry experiment to determine the formula of tin iodide. The miniature kit for the course on *Our Chemical Environment* referred to in Subsection 3.1 supports a varied set of experiments, including an investigation of the role of pectin in the 'setting' of fruit jams, jellies or conserves, the extraction of nicotine from tobacco, and the analysis of minerals.

Kits for biology courses have included an even wider range of items – for example an ultraviolet moth trap (for a classification exercise, the moths being subsequently released), live organisms as noted above, a model of the human brain, enzymes and seeds. Some experiments require students to collect their own material. For instance a Foundation level biology experiment has involved students in an investigation of mortality factors based on an analysis of leaves attacked by the holly leaf miner. Students find the leaves themselves (holly having been chosen for this exercise because it is well distributed through the UK), but the hand lens required to examine them is sent in the kit. Experiments devised for second level students have covered the respiration of blowfly maggots, membrane permeability, the effect of a plant growth regulator on the growth of cucumber seedlings, the genetics of bread mould, and animal and plant histology (these last requiring the supply of a microscope and the preparation and interpretation of slides).

For large population courses, the collation of data from individual students can be of great interest. Simple observations made in the 1980s by Foundation course students on the relative proportions of normal and melanistic forms of the peppered moth (*Biston betularia*), were assembled into a national survey, with the results reported back to students through a TV programme. The national distribution of the two forms could then be clearly seen to be linked to environmental factors. On other courses similar surveys have been carried out of concentrations of atmospheric sulphur dioxide and of water quality. These kind of surveys make positive capital of the geographically scattered distribution of UKOU students, and can make a genuine contribution to scientific research by collecting data on a national scale.

Reference has already been made to the HEK for the physics course *Images and Information*, which includes all the apparatus for a practical holography experiment (laser, optical bench, camera, photometer and monobath developer; students need only to supply the film). The hologram itself is then submitted to the student's tutor as part of the assessed coursework.

Within the home experiment framework, it is also possible for more advanced students to undertake self-directed project work. For example, third level ecology students are expected to complete and write up a 40-hour, normally field-based, practical project, for which certain types of equipment commonly used for biological fieldwork (quadrat, quadrat point, binocular microscope, pooter) are available from the home kits warehouse on request. Questions that students on this course have investigated are many and diverse – for example, the influence of salinity on the distribution of plants in estuarine waters, the factors affecting the recolonization of abandoned industrial sites, the variation of animal population in leaf litter between one woodland and another, and the recovery of heathlands from fire damage. One of the major difficulties of project work for remote students is access to expert comment and 'supervision' of the type required for most novice researchers to achieve satisfactory results and to learn from the experience. In the ecology course, this problem is addressed by getting each student to submit the work for assessment in two stages. The first stage consists of an outline of the proposed project. A local tutor with a background of ecological field work then supplies detailed feedback on these initial plans, and is accessible for further guidance during the experimental phase. This early contact has been found to be crucial in honing projects so that they are likely to produce valid results within the limited timescale available: enthusiastic but inexperienced students usually underestimate the time required to complete field work, and need the pragmatic advice of an experienced researcher in order to focus more clearly on their aims for the project, and/or to reconsider their working methods. Once the practical work has been carried out and analysed, the student produces a final report, which is submitted to a different tutor so it can be assessed purely on its merits as a research account, without knowledge of the previous negotiation between student and 'supervising' tutor. It appears that the students on this course have greatly valued the free choice of subject, which gives them a sense of 'ownership' of the project, and allows them to pick a topic according to their individual circumstances and special interests, and with an eye to the habitats available locally. 'Personalized' projects also permit students with special needs to undertake feasible but valid project work: for example, wheelchair-bound students can research accessible ecosystems such as gardens. It is interesting to note that the project element of this course is backed by a range of courseware: textual materials covering field work techniques and methods of statistical analysis, a video illustrating how topics may be chosen, and a residential week at a field studies centre.

3.3. Designing home experiments

Because home experiment kits have been a feature of the science provision of the UKOU since the earliest days of the institution, some of the kits have gone through several generations of development: for example, the Science Foundation course has already been completely rewritten three times. Survey data on student perception and usage rates of these kits has been collected in nearly every case. During this twenty year evolution, it has gradually become apparent that, apart from a consideration of specific learning outcomes, the major issues to be addressed when designing a home experiment are the following.

Time and logistics

The early course teams soon found that their initial estimates of the amount of time students would require to complete home experiments tended to be gross underestimates (often by a factor of two). During the production of subsequent courses, home experiments were therefore subjected to iterative developmental testing procedures. It was also discovered that, although some initial testing (e.g. to ensure satisfactory operation of items of equipment) may usefully be carried out by expert course team members, the identification of many potential problems and realistic timings come only when prototype experiments are attempted by people with academic and practical backgrounds similar to those of typical students.

All other aspects of HEK design are in a sense subservient to the basic requirements that it must be not only safe but also practicable for isolated distance learners to carry out the experiments in their own homes. They cannot be expected to clear more than a typical kitchen table's worth of space and even that will probably have to be restored to its normal function in time for the next meal. Setting up, and clearing up, times must always be taken into account in estimating the total time required for carrying out an experiment.

Objective setting

Because of time pressures and the physical inconvenience of clearing a suitable workspace, unpacking the kit items and setting up an experiment, students have to be convinced of the value and relevance of a home experiment before they will even attempt it. They will do an experiment if an important course objective can be achieved in no other way. Conversely, if the associated objective can be fulfilled simply by reading about an experiment, then many students will not bother to try it for themselves. Course designers may, for sound educational reasons, wish to expose students to practical demonstrations of phenomena described in the texts, or to practical illustrations of particular principles; such activities can undoubtedly have a place in the overall course structure, but it is now generally recognized that this place is not the home experiment component. (It will be suggested later in this chapter that televisual media are probably the most effective delivery systems for demonstrations and illustrations.)

Integration with other course material

Another way of minimizing student avoidance of practical work is by ensuring that the home experiments are fully integrated into the main coursework, not an adjunct to it. In accordance with this strategy, there is an increasing trend for home experiment instructions to be written in to UKOU science courses at appropriate points in the main texts (which carry the bulk of the instruction), rather than being issued in the form of supplementary sheets of paper.

Links to the assessment strategy

A third method of reducing 'experiment evasion' and emphasizing the importance course designers attach to a particular experiment, is to make HEK-based practical work part of the assessed coursework. As will be discussed in Chapter 11, surveys have shown that 'the next assignment' is the single most important factor in pacing most students' work. A non-assessed experiment will be regarded at best as optional and at worst as irrelevant.

Safety considerations

Even if an experiment satisfies all the criteria listed above, it may still not be suitable as a *home* experiment because it cannot be made sufficiently safe. For novice students working alone and/or in a domestic environment, the safety constraints are particularly severe. Not only will there be problems with fume extraction, and the disposal of waste chemical products, but any accidental spillage may occur on an expensive kitchen work surface rather than a tiled laboratory bench. These types of considerations are the main driving force towards the trend to small scale chemistry experiments noted in Subsection 3.1. Electrical equipment must meet the same safety standards as domestic appliances. All possible contingencies must be covered in the instructions issued to students.

User-friendliness

Many inexperienced students are quite nervous about doing experiments, so user-friendliness, in relation to both the apparatus and the manual, is vital. In devising home experiments, instructors should always be conscious of the fact that students will have no immediate access to expert help in case of difficulty. Admittedly this applies to all components in a distance teaching system, but the lack of an expert troubleshooter may be felt particularly acutely in the context of practical work. Equipment designed for home usage must therefore be exceptionally reliable. As Greenfield has pointed out,

> students' attitude to experimental work can be affected by the performance of their instruments. Unfortunately, naive users are not in a position to know whether the 'fault' is theirs or that of the equipment, if things go wrong.

(Greenfield, 1984, p. 114)

Nothing can be taken for granted. The manipulative processes involved in carrying out the experiment have to be clearly explained to the student. Figure 10.1 shows simple examples of the use of a diagram for such a purpose; more complicated manipulations might have to be demonstrated on a video.

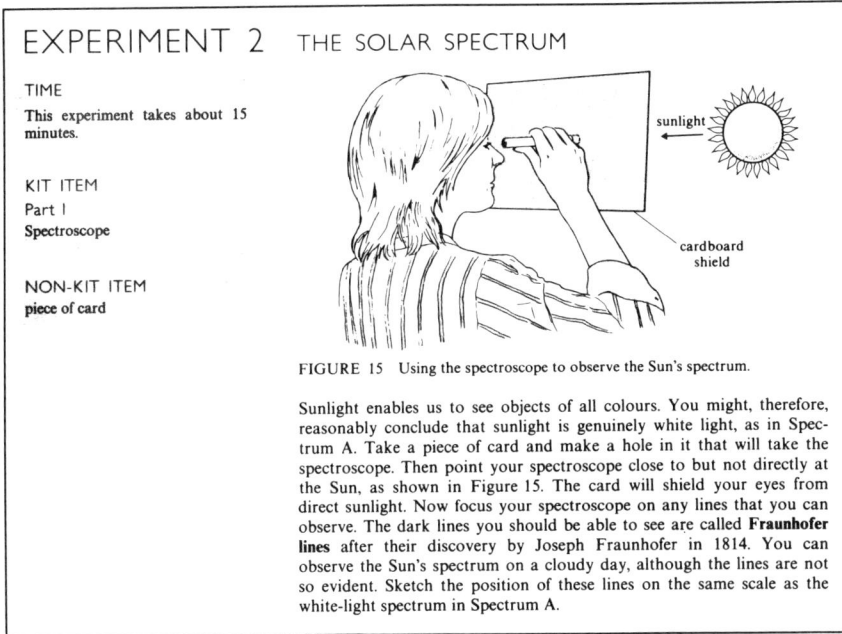

Atomic spectra, being characteristic of the elements that produce them, provide a means of analytical identification of elements. You will meet atomic spectroscopy again in Units 13–14 and we illustrate its use in the TV programme 'Steel, stars and spectra'.

EXPERIMENT 2   THE SOLAR SPECTRUM

TIME
This experiment takes about 15 minutes.

KIT ITEM
Part I
Spectroscope

NON-KIT ITEM
piece of card

FIGURE 15   Using the spectroscope to observe the Sun's spectrum.

sunlight

cardboard shield

Sunlight enables us to see objects of all colours. You might, therefore, reasonably conclude that sunlight is genuinely white light, as in Spectrum A. Take a piece of card and make a hole in it that will take the spectroscope. Then point your spectroscope close to but not directly at the Sun, as shown in Figure 15. The card will shield your eyes from direct sunlight. Now focus your spectroscope on any lines that you can observe. The dark lines you should be able to see are called **Fraunhofer lines** after their discovery by Joseph Fraunhofer in 1814. You can observe the Sun's spectrum on a cloudy day, although the lines are not so evident. Sketch the position of these lines on the same scale as the white-light spectrum in Spectrum A.

Figure 10.1. Diagram from a foundation course text illustrating the use of a hand spectroscope (Open University, S102, Units 11/2, p. 26).

Instructions have to be broken down into a series of small steps and must be unambiguous. It is important to ensure that students are not frustrated (or, worse still, endangered) by an inability to follow the experimental protocol, or by a failure to achieve a result at the end. For these reasons the vast majority of science home experiments are 'prescriptive', rather that open-ended, although there are exceptions, some of which have been noted above in relation to project work. An example of the instructions for a third level chemistry home experiment is shown in Figure 10.2. It is interesting to note, on both Figures 10.1 and 10.2, the estimation of time required to complete the experiment and the details of the items needed.

It is sometimes possible to capitalize on the multiple media approach. For example, it may be helpful to put instructions on to audio-tape instead of presenting them as a printed 'recipe'. One advantage of this media mix is that spoken explanations can often make the process seem less formal and intimidating. Another is that students can act upon spoken instruction without having to take their eye away from the microscope, put down one of the two rock samples they are comparing, or stop stirring the solution. Mixed-media approaches are also very valuable in the pursuit of more complex objectives. A third level UKOU project in organic chemistry, which requires students to evaluate a proposal for synthesis, features thin-layer chromatography carried out by the student together with video demonstrations of high performance liquid chromatography.

─────────────── E X P E R I M E N T 2 ───────────────

APPARATUS

dropping pipettes
10 cm³ measuring
cylinders (2)
microspatula
pH paper
rubber stoppers for
test-tubes
standard spatula
test-tubes and rack

CHEMICALS

citric acid
common salt
product from Experiment 1
0.5 mol l⁻¹ sodium
hydroxide solution
1,1,1-trichloroethane
(TCE)

ISOMERIZATION OF CYCLO-*SS*-ISOLEUCYL-ISOLEUCINE WITH SODIUM HYDROXIDE

This experiment should take about 45 min, including 10 min reaction time, which could be usefully employed preparing for the next experiment.

Dissolve 1 microspatula-end of your recrystallized product in 1 cm³ of IMS in a test-tube*. Add 2 cm³ of 0.5 mol l⁻¹ sodium hydroxide and allow to stand for about 10 min (*Note 1*). Add citric acid to the solution until just acid (pH paper, 1–3; add 1 standard spatula-end at a time and check with pH paper after each addition) and then 1 standard spatula-end of common salt followed by 1 cm³ of TCE. Insert a rubber stopper and shake the test-tube for one minute (not too vigorously). Allow the layers to separate and, by inserting a dropping pipette to the bottom of the tube, carefully withdraw most of the lower layer (TCE) (being careful not to pick up droplets of the aqueous layer), and transfer it to another test-tube and stopper.

Note 1 The literature procedure suggests dissolving the cyclic dipeptide directly in aqueous sodium hydroxide. The procedure suggested here saves time and is just as effective. Ignore any slight precipitation that may occur in this experiment.

* *You can reduce significantly the time required to prepare a solution by inserting in very hot water, the end of the test-tube containing the components.*

Figure 10.2. Instructions for a home experiment carried out by students on a third level organic chemistry course as part of a project associated with high performance liquid chromatography, and stereoisomers of cycloisoleucylisoleucine (Open University, *Organic Chemistry: A Synthesis Approach*, S344, Block 2, pp. 27–8).

3.4. Costing home experiment kits

There is no denying that HEKs are expensive and constitute a direct add-on cost to the course.

Development costs are well in excess of those for laboratory work in conventional institutions. Academic design of a kit needs to proceed in parallel with the development of other course components. If, as in the UKOU, the production course team works for two or three years, the investment of academic time is considerable. However, development costs can to an extent be offset against high student populations and extended course life: the UKOU registers in excess of 3,000 students per year on its Science Foundation course and each version of this course is expected to be presented annually for eight to ten years.

Capital costs are also high. In conventional laboratory-based classes, students usually work in pairs and often rotate round a series of experiments. Ten sets of equipment for each of four experiments is all that is required for a class of 80 students. Under an HEK system, every one of those 80 students would require a full set of equipment for each experiment.

Packaging costs are often non-trivial, especially if trays have to be specially designed and manufactured in order to send fragile items through postal or carrier services.

Maintenance and refurbishment costs for non-disposable kits represent a major commitment for the lifetime of the course. Kit items returned by one year's students have to be checked and if necessary serviced, recalibrated or repaired before being issued to the following year's cohort. Consumables must be replaced annually.

4. TELEVISUAL MEDIA

As already noted in Chapter 7, there are some types of practical work, the most obvious of which is the demonstration experiment, for which a televisual presentation is ideal. The educational advantages of the demonstration experiment are well rehearsed. For instance:

- Demonstrations are an effective way of tying together theory and phenomena. The demonstrator can order the argument, stress the relevant features, and suppress the kinds of irrelevancies that may sidetrack students when they carry out experiments themselves.
- Demonstrations set standards of good practice. Students can see how a professional scientist handles particular items of apparatus, and how a well-performed experiment proceeds.
- The performance of the single demonstration, as opposed to parallel experiments carried by large numbers of students working individually

or in small groups, greatly reduces the amount of material used. This is especially important when the experiment involves expensive consumables, hazardous chemicals that are difficult to dispose of safely, live animals, plant or animal material for dissection or material collected from the natural environment (such as rock samples or fossils).

The filmed demonstration can on occasion have additional advantages over an experiment performed 'live' in front of a student audience. For example:

- Pre-recorded demonstrations are not going to go wrong! The students can see the experiment presented free of other distractions.
- Techniques such as zooming, freeze-frame and split screen can focus attention on the most crucial aspects of the demonstration.
- Specially shot sequences for TV or video allow students to see difficult or downright dangerous experiments that could never actually be carried out in the classroom setting. A much praised TV programme for the UKOU Science Foundation course, designed to provide experimental support for conceptual material in the periodic table of the elements, had as one of its core sequences a demonstration of the reaction of alkali metals with water. Starting with the already vigorous reaction of lithium, the experimenter was seen to retreat behind successively more elaborate screening as he proceeded to the increasingly violent reactions of the heavier elements. By the time he arrived at caesium, the reaction was explosive. His summing-up line 'As you can see, things become more and more terrifying as I go down the Group' became almost a catchphrase among students of the course, illustrating the considerable impact that the dramatic experiment can have in stimulating interest and activating the memory.

In helping students to achieve practically oriented objectives, the use of televisual media is not confined to the presentation of experiments. Through the television, the remote student can also be given access to other types of practical experience, such as 'visits' to field locations, astrophysical observatories and major research facilities – at the other side of the world if that is appropriate. Vicarious experience of field work can in fact score over the real thing in a number of significant ways. Students can be 'taken to' distant locations and hostile environments (e.g. polar regions, the high Himalayas or the deep ocean), which would be quite inaccessible to a normal field party. Through the TV camera, every student can see the particular features being highlighted by the presenter, and study them in comfort, whereas in the situation of a real field trip the problems associated with groups of students working in difficult terrain or unpleasant weather conditions may mean that some students actually miss key points.

For active student involment, as opposed to demonstrations, videocassettes are more versatile than TV transmission. As discussed in Chapter 7, a broadcast TV programme is scheduled and ephemeral, and tends to be watched passively, whereas a videotape can be viewed repeatedly at the student's convenience. A video sequence can therefore be tightly integrated with textual or other components of the course, and students can be required to interact actively with it. A video 'snippet' is also the ideal way of showing students the correct way to handle a piece of equipment. Used in conjunction with a HEK, it can offer a valuable method for the development of manipulative skills and techniques.

5. COMPUTER MEDIATED SYSTEMS

The use of computer programs to simulate experiments, and the educational advantages and drawbacks of such simulations, have already been discussed at length in Chapter 8. Even more 'realistically', as mentioned in Chapter 9, multimedia packages, such as interactive videodisc (IV), CD-I and CD-ROM systems, offer unrivalled flexibility as simulated laboratory or field environments. Students can watch a sequence of an experiment in progress and take readings from instruments shown on the screen far more easily than from videotape presentation of similar material. Furthermore, because of the computer control, more sophisticated measurements become straightforward: using a mouse, a trackerball, or a suitable cursor device, co-ordinates can be taken directly from the screen. The control computer can also log the results, facilitating inputs for data handling and analysis. The 'virtual laboratory' can be better equipped that any traditional teaching lab, since the one set of expensive apparatus used in setting up the package then becomes simultaneously available to all students. The advantages of safety and limited usage of consumables, already discussed in Section 4 with reference to filmed demonstrations, apply equally to multimedia presentations but the interactivity adds an additional dimension, allowing students to explore the consequences of alternative strategies without risk. These features open up a huge range of possibilities for the teaching of practical skills, and the versatility of multimedia packages is such that they can be designed so as to provide students with practice in many of the higher order and hybrid skills. Examples of such uses (see e.g. Ross (1991) and references therein) include:

- a visual database on videodisc, from which students can make measurements and undertake analysis of various types of motion, both from demonstration experiments and real-life situations such as athletes in action and ball games (Glover *et al.*, 1989). Exercises of this type address basic practical skills of accurate measurement, estimation, data handling, error analysis, etc.
- simulated field trips, which allow students to discover otherwise inaccessible environments. An IV program developed by the Dutch Open Universiteit (Beijderwellen, 1990) allows geology students to 'visit' a number of different locations, to correlate their observations with

geological maps, rock samples, fossils and field experiments and ultimately to reconstruct the geological history of the area. This kind of substitute for field experience is invaluable for remote learners, who may be without access to many field locations, and/or unable to meet with knowledgeable guides to a given area. Similar vicarious field trip programs are currently being developed for CD-ROM, as described in Chapter 9.

- replacement of some laboratory sessions by a 'virtual laboratory', in which, in addition to the advantages of safety and use of consumables noted above, offers students immediate integration of practical work with reference material, animations, stills, video clips of applications, context-specific tutorial help and self-test, on-line, quizzes. Jones (1988) has described an IV chemistry laboratory which was well received by undergraduates in the mainstream sector and which evaluation showed to be effective not only in developing their practical skills, but also in consolidating their conceptual understanding. F. De Vries and R. Niesink of the Open Universiteit, The Netherlands (private communication, 1991) found IV to be an efficient medium for giving remote students of toxicology vicarious experience of practical work, and also converted the material for CD-I.

6. LABORATORY CLASSES

Although many of the objectives associated with practical work can be attained through demonstration or simulation, and manipulative skills can be developed by experiments that students carry out unsupervised at home, there are nevertheless certain kinds of experience that students can acquire in no other way than by working in a well-equipped laboratory environment. Science graduates are expected to be familiar with the operation of instruments – such as mass spectrometers or gas chromatographs, say – that could never be included in a HEK kit. They may also need to have performed experiments requiring liquid nitrogen temperatures, a high voltage supply, or a fume cupboard. To provide these sorts of laboratory experience, and/or the chance to do work in the field, many distance education science courses include a compulsory day or residential school component. UKOU students attend a university campus (or field studies centre), during that institution's vacation period, making use of halls of residence, social facilities and undergraduate laboratories.

There is little that need be said here about the actual experiments or field work carried out at residential schools, since they are directly comparable to those performed by students in the mainstream sector. However, this kind of practical experience is bought at a price. People who have chosen distance education in order to balance study with the competing demands of jobs and family commitments have to pay travel and accommodation costs for residential schools, take leave from work, arrange for the care of dependents. Attendance is particularly difficult for some groups, such as those with disabilities, single parents, and women from certain religious or cultural backgrounds. Given these problems, and the fact that laboratory and field-based time for distance education students tends to be very limited, it is especially important that it be well used, with particular emphasis on objectives that cannot be achieved in other settings.

7. CONCLUSION

Early suspicion over the extent to which the distance education sector could achieve the learning outcomes traditionally associated with practically orientated objectives is now being replaced by a realization that although the instructional strategies required to train remote students in practical scientific skills may be different to those applied in laboratory or field-based classes, they need be no less effective. Indeed there is a growing recognition in the wider science-teaching community that many of the aims of traditional instruction in practical techniques should certainly be subjected to more careful scrutiny and might even be better achieved by other means.

In designing course programmes and components, science educators in distance teaching institutions often have at their disposal a wider range of media than their counterparts in the mainstream sector. Because of this flexibility, some of the strategies that have first emerged in the distance education context arguably have superior outcomes to those achieved in conventional laboratory classes. For example, one real bonus of the multiple media distance education package is its flexibility, which allows tight integration between theory and practice. A demonstration experiment viewed on videotape just at the time the student is working on the related theory has immediacy and relevance. An opportunity to carry out a brief exploratory investigation, or to get a feel for a phenomenon in a real-world context, can be an invaluable prelude to a discussion of the underlying theories. Such integration between the practical and theoretical sides of the subject is further fostered in distance education courses by the instructional design framework which is imposed on the whole package of learning materials, and is far more difficult to achieve within the traditional pattern of separately timetabled and independently supervised lectures and practical classes.

11

ASSESSMENT

If we wish to discover the truth about an educational system, we must look into its assessment procedures. What student qualities and achievements and activities are actively valued and rewarded by the system? How are its purposes and intentions? ... The answers to such questions are to be found in what the system requires the student to do in order to survive and prosper. The spirit and style of student assessment defines the de facto *curriculum.*

(Rowntree, 1977, p. 1)

1. INTRODUCTION

In this chapter we describe the processes involved in assessment, dealing with the problems of assessment at a distance, the difference between formative and summative assessment, and the different strategies involved in the design of tutor- and computer-marked assignments and examinations. The particular features of assessment in science education will be discussed by illustrating the type of assessments in use. As well as grading students, the assessment component of a course can provide useful information to be fed back to the course designers which can help them evaluate their teaching, but the discussion of assessment in relation to evaluation will be delayed until Chapter 14.

A useful wide definition of assessment was given in the report of the Task Group set up to examine assessment in English schools (Department of Education and Science, 1988), as follows:

A general term embracing all methods customarily used to appraise performance of an individual pupil or a group. It may refer to a broad appraisal including many sources of evidence and many aspects of a pupil's knowledge, understanding, skills and attitudes; or to a particular occasion or instrument. An assessment instrument may be any method or procedure, formal or informal, for producing information about pupils: e.g. a written test paper, an interview schedule, a measurement task using equipment, a class quiz.

(DES, 1988, Preface and Glossary)

This quotation emphasizes that the word assessment is used to describe many different types of activity. This also becomes clear when one considers the variety of purposes to which assessment is put. Assessment is sometimes just equated to the method by which students can be graded, but grading is only one possible outcome of assessment. Other vital purposes include diagnosis, both in self-assessment by which the students themselves assess their progress on a course, and in assessment used to provide part of an evaluation of teaching materials. The most important role of assessment is finding out about the student's learning and using the information gained for teaching purposes. A teacher needs to engage in diagnostic assessment, assessing a student's developing strengths and weaknesses in order to plan effectively for the next stage of learning. Students need to have information on their performance while studying, and Rowntree asserts that

> Feedback, or 'knowledge of results' is the lifeblood of learning . . . Effective feedback enables the student to identify his strengths and weaknesses and shows him how to improve where weak or build upon what he does best.
>
> (Rowntree, 1977, p. 24)

There is evidence to suggest that detailed feedback on performance can improve students' learning. For example, giving students *individualized* comments on their work seems to help (Page, 1958). We will develop this point more fully in Chapter 13 in relation to correspondence tuition.

Assessment is important for students' learning not just because the students (and their teachers) find out what they know about the course being studied, but because for many students the course is actually defined by the assessment. Becker, Geer and Hughes (1968) describe the reality of the dominating effect that grading procedures have on American college life, while Miller and Parlett (1973) in their study of British university students found that almost all the students they interviewed took an instrumental approach. One of the undergraduates involved in their study said

> What is the purpose of the examination game? . . . You know you want to get a certain class of degree, within the system, but as far as assimilating knowledge properly is concerned, it just doesn't work, because if you play the game properly you're choosing all the time
>
> (Miller and Parlett, 1973, p. 53)

Both examination practices and continuous assessment can influence what students perceive to be 'playing the game'. Rowntree notes how a perspective focused on grade-point averages can influence students' attitudes and behaviours:

> George Miller (1976) gives an illustration of how this instrumental preoccupation worked out in a medical school when the Pharmacology department found itself with falling attendance at its Thursday afternoon laboratory sessions because the department of Physiology had instituted

a weekly quiz on Fridays. Pharmacology successfully retaliated with laboratory examinations on Thursdays!

(Rowntree, 1977, p. 125)

2. ASSESSMENT: TYPOLOGY AND VALIDATION

Assessment may be classified in a number of different ways, but the most important categories relate to the purposes of the assessment and to the reference standards.

Formative/summative assessment

Summative assessment is designed to provide information about what students have already achieved. In contrast, formative assessment is designed to provide information to students about their progress in learning. Also, the results of formative assessment can be useful to teachers or instructors in planning future learning experiences. In practice, the term summative is often used to denote assessment that is graded and counts towards the students' course credit, while formative assessment is either not graded or graded only for the students' information.

Norm referenced, criterion referenced and self referenced assessment

The norm referencing view of testing assumes that performance on a 'good' test would be a normal distribution of scores (in the statistical sense). The aim of such tests would be to indicate how students perform in comparison to one another. In contrast, criterion referenced tests assess students' performance against some fixed standards set by the teacher or instructor. In self referenced assessments the student's current performance is compared with his or her performance on a previous test, so giving a measure of individual progress.

One of the most important considerations in designing tests relates to their validity and reliability.

In any assessment system there are two key questions: Does the result truly reflect the students' achievements? Is the result a fluke?

The answer to the first question determines whether or not the assessment is valid, and requires for example the judgement of a subject specialist to assess whether the scoring responses to the test are fair and correct. The second question can only be answered by trying to assess reliability, i.e. whether a similar assessment would produce a similar result on a different occasion. There are statistical techniques that can be used to determine reliability. For example, one way of judging this is to split a test in half and calculate a correlation coefficient between a student's performance on each part of the test (i.e. split half reliability). However, other affective factors can influence students' performance on assessments, not least of which is the students' perception about the relevance and importance of the tasks set. For example, Laurillard (1979) found that an assignment was tackled differently according to how students perceived its importance and interest. Other components of assessment systems, especially those such as projects and examinations which can be individually very important, must also be validated and there is a considerable body of research on their fairness.

Practical reports and projects

Assessing students' practical work by means of written reports is fairly common. This method is often quite subjective because, as with other written reports, 'such variables as neatness, writing skills, volume and degree of completeness can bias evaluation' (Giddings, Hofstein and Lunetta, 1988, p. 168). Also written reports do not provide direct evidence of students' practical skills as displayed in the laboratory. When attempts are made to assess such skills directly using independent observers, as in some school examination boards' practical examinations, problems still arise. Bryce and Robinson (1985) report on the use of a technique whereby pupils circulate round a small number of experiments while being observed. Linn and Songer (1988) report on the 'computer as lab partner' project at Berkeley where the computer was used in a variety of ways to enhance and simultaneously keep records of students' performance in the laboratory.

Bradley (1984) hypothesized that personal supervisors of undergraduates would be able to be unbiased in their assessment of students' projects. She predicted that second markers furnished only with the names of students would mark males more extremely than their supervisors did, i.e. males would be awarded more firsts and thirds than females. She found that in four university departments this happened.

Examinations

Many students find it a source of worry that they have to sit a final examination. Part of their worry must come from such well-known pieces of research (such as Hartog and Rhodes (1936)) on the potential for unreliability in grading. Final year students at Edinburgh University were interviewed by Entwistle and Entwistle (1991) and asked about their revision strategies and about their attempts to develop understanding of material to be learned. Findings from this study question whether traditional examinations test deep conceptual learning. Some courses such as an UKOU certificate course for science in the primary curriculum have pioneered the approach that the grade for a substantial piece of project work can replace the final examination.

Final degree results in a number of universities have come in for detailed scrutiny. One issue that has interested examiners is whether sources of gender bias have been eradicated, judged by the overall pattern of results. Mason (1985) found no overall differences in the overall achievement of male and female students in a study of the UKOU students' degree success in science. Of course one advantage that distance education examination systems may have is that students are usually anonymous to their examiners. Belsey (1988) looked at degree performance in the Arts Faculty at University College Cardiff before and after 1985, when blind marking was introduced. Women achieved a significantly higher proportion of good degrees in English (i.e. firsts and upper seconds) relative to men after 1985. Goodheart (1992) has reviewed exam performance at the University of Cambridge, UK and Rowell (1991) at the University of California at Berkeley. Rowell found no significant differences in the final grades achieved by male and female students at Berkeley as compared to Goodheart's findings that women in Cambridge achieved fewer firsts and thirds. There are however differences in both the admissions and examination systems. At Berkeley students take examinations at the end of each semester and accumulate a grade point average. Rowell's hypothesis was that this was a consequence of the fact that at Berkeley 'less elegant' examinations were used (i.e. exams without essay type responses but more use of short answer and multiple choice formats). A fascinating discussion of the extent to which equity is possible in assessment situations is given in Gipps and Murphy (1994).

3. ASSESSMENT OF SCIENCE LEARNERS

Arguably, the most interesting writing about assessment of science learners in Britain came out of the Assessment of Performance Unit (APU). This was set up in 1975 to provide information about the national levels of performance of schoolchildren in a number of areas of the curriculum, one of which was science. The main aim of the APU was 'to produce and make generally available national pictures of performance, not to report on the performance of individual children, schools or local education authorities'. Science was to be assessed at three ages: 11, 13 and 15 years. The project began by establishing a framework as a basis for assessment. This framework was process-based and assessment procedures were designed to monitor children's abilities to perform scientifically. The categories of performance were as follows: use of symbolic representation, use of apparatus and measuring instruments, observation, interpretation and application, design and performance of investigations. As a result of its investigations, the APU

Science group has published a number of reports. Harlen (1992) describes the overall findings on pupils' performance as follows: that children were well able to use skills related to observation and measurement and were able to set out practical problems in an appropriate way and with signs of real interest – skills generally applicable across the curriculum. However, for skills specifically related to science performance (recognition of patterns, explanation of events using science concepts) there was generally lower performance.

As discussed in earlier chapters, Millar (1989) has drawn attention to the difficulties inherent in measuring process skills and has written critically of unwarranted assumptions about the ease of doing so, questioning the feasibility of teaching processes as content-independent strategies, of testing them in related tasks which may be different in context and content, and of distinguishing levels in such tasks.

Compared to the problems associated with assessing skills, the task of assessing students' knowledge of a concept might seem simple. Here too, however, there are several problems.

> The relative failure of pupils to apply science concepts must be a cause for concern to all engaged in science education. Why is performance so low? ... Could it be that science educators value only that knowledge and understanding which is inherently difficult?
>
> (DES, 1985, p. 38)

The literature on students' understanding of science concepts is full of evidence that although students can pass examinations, their understanding of very basic concepts is incomplete (see for example the work by Di Sessa, Driver and Gilbert referred to in Chapter 3).

One feature which stands out from all this research is that question setters need to pay attention to the context dependency of students' knowledge which can make meaningful tests of students' conceptual knowledge difficult to construct (see e.g. Engel-Clough and Driver, 1986; Nuttall, 1987). There are also clear gender differences in science performance which some researchers have linked to the way the sexes respond differently to different contexts in science (Johnson and Murphy, 1986; Murphy, 1991).

4. ASSESSMENT IN DISTANCE EDUCATION

The formal assessment of student achievement, the provision of feedback which students need to judge their performance, and a measure of the teaching effectiveness of the course, are all requirements that the assessment component of a distance teaching system should provide. The more informal channels through which some of these purposes are achieved in face-to-face tuition are not available, and self-assessment exercises in the course material, while valuable, cannot provide individualised feedback. Instructional theory suggests that personalized feedback on performance is extremely important in aiding and reinforcing learning; in a distance education system such feedback can only be provided equitably by building it in to the assessment and marking procedures.

The main types of assessment used in distance education are computer-marked (e.g. multiple choice) assignments, tutor (i.e. script)-marked assignments, both of which the students can complete at home, without time constraints and in open book mode, and examinations, which are held under controlled conditions. 'On line' computer assessment may be used more extensively in the future as electronic links and computer conferencing facilities become more widely available. Continuous assessment and examination grades can be conflated to give an overall score. However, combining different forms of assessment in this way is not without its problems.

The results of continuous and final assessment often do not agree; indeed, in some cases exam results may flatly contradict the evidence from continuous assessment. Most commonly the student may have poor examination technique and may work better with unlimited time and access to textbooks. Sometimes the two components may be testing different objectives, or different cognitive levels of objectives. If the marking of continuous assessment is done by different people from those who mark the exam, the formative aspects of the continuous assessment can be overstressed, resulting in students gaining a false sense of security about their performance. Equally, final examinations questions can be set without checking that students have been given adequate opportunities to develop and practise the skills which are tested. In working on assignments, remote students may have access to other sources of help, friends or relations. In examinations, the result is definitely their own effort.

Given the disparity in marks that can arise between assignments done at home and examinations, it is often considered desirable to build another different form of assessment into the system. One option, which the distance education sector has borrowed from the mainstream, is to require a viva voce examination in cases where there is a very large discrepancy between a student's continous assessment and examination scores. Another option, for courses with a compulsory day-school or residential component, is to get tutors to give marks for performance and progress at these events. The UKOU adopted this system initially, especially for courses with a compulsory laboratory element, as it was thought to be comparatively easy to give marks for students' approach to practical work. However, the idea gradually fell into disfavour, partly because the subjective nature of the assessment meant that the scores were difficult to standardize, and partly because the grading function of the session was felt to get in the way of its (more important) teaching and learning functions. A hybrid form of assessment that combines face-to-face tuition in the laboratory with the later submission of an assignment on the lab work can overcome some of these difficulties, but it is often awkward to administer in practice.

Integrating assessment into course design

Placing emphasis on assessment at an early stage of course design is very important. It ensures that instructors clarify for themselves at an early stage the relationship between the assessment and the overall aims and objectives of the course. It also allows the course designers to get a better idea of the overall workload of the course, as the assessment itself will be a significant part of that workload (see e.g. Chambers, 1992). A course that contains too many assessments or individual assignments that take too long can be demotivating for students. It is easy to underestimate the time that students can spend on a particular assignment:

> It took me at least 24 hours that first TMA [tutor-marked assignment] question but it came back with a good mark and I was very pleased, though I can't always spend that much time on one question.
> (Lunneborg, 1994, p. 76)

Tan (1992) describes the 'profound negative influence' of frequent summative assessment on students' conceptual understanding. Overall, it is usually preferable that the original course designers should prepare a bank of assessment material that can be used for each subsequent presentation of the course, and this reduces the risk of a drift in standard with time. On the other hand, in a system in which the course materials remain fixed for many years, a change in assessment policy may be the only way to amend the course in the light of feedback.

5. DESIGNING ASSESSMENT INSTRUMENTS FOR DISTANCE EDUCATION SCIENCE COURSES

In this section we present criteria for tests used in science courses, and illustrate these with examples from UKOU assessment items, beginning with the main instruments for continuous assessment: computer-marked assignments and tutor-marked assignments.

Computer-marked assessment

Computer-marked assignments (CMAs) are a quick and efficient way of providing students with information about their progress. Figure 11.1 gives an example of one of these. The use of CMAs presupposes that it is possible

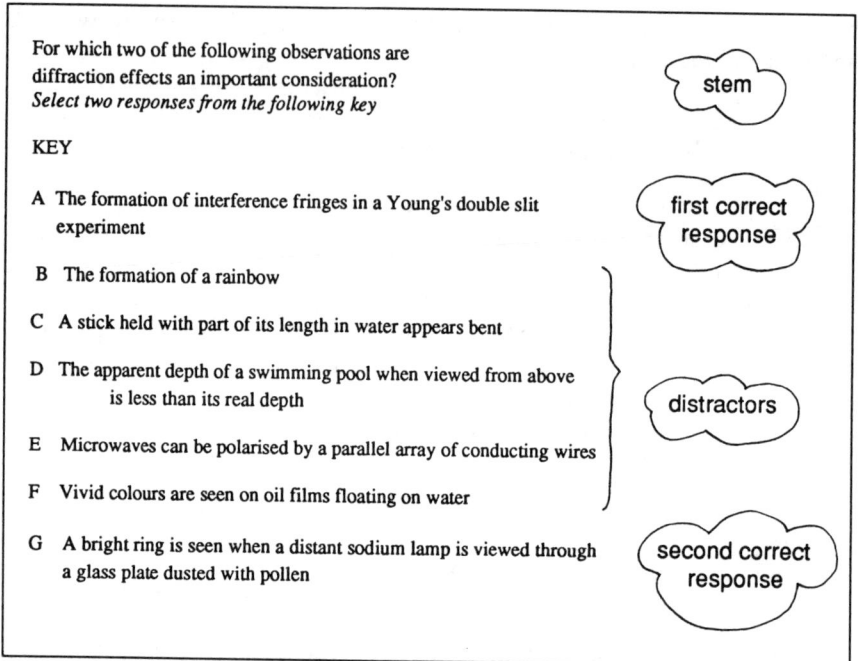

For which two of the following observations are
diffraction effects an important consideration?
Select two responses from the following key

KEY

A The formation of interference fringes in a Young's double slit
experiment

B The formation of a rainbow

C A stick held with part of its length in water appears bent

D The apparent depth of a swimming pool when viewed from above
is less than its real depth

E Microwaves can be polarised by a parallel array of conducting wires

F Vivid colours are seen on oil films floating on water

G A bright ring is seen when a distant sodium lamp is viewed through
a glass plate dusted with pollen

Figure 11.1. Structure of a foundation level computer-marked assignment question (Open University, S100, CMA 54, Q.V1, 1974, with additional labelling).

to ask various questions on a topic for which there is only one correct answer or set of answers, so that assignments can be constructed which involve the learner in making a selection of the correct answer from a list of possibilities. When the students' responses are recorded on a standardized form they can be optically scanned and passed into the computer. The computer is programmed with information about which are the correct answers and the weighting factor of each correct answer, so that a final grade can be calculated. The computer can also provide developmental feedback from summaries of overall student performance on a particular question.

It is of course technically feasible, and educationally desirable, for computer-marked tests to be taken by students 'on line', so providing instant feedback to the learner. At present, such schemes are not in widespread use in distance education, due to problems of access, but are likely to develop quickly in the future as institutions move towards arrangements for the electronic campus or virtual university (see Chapter 15).

One of the strong arguments for CMAs lies in the control which can be exercised over their setting by comparison with the relative lack of control over the marking process which occurs with other forms of assessment. Another is that CMAs allow a far wider coverage of the subject than is usually possible with other types of assignment. A few long written answers may not reveal the gaps in a student's knowledge, while a well constructed CMA could do so quite quickly.

A multiple choice question is often referred to an item. Figure 11.1 shows an example CMA item with its various parts distinguished – stem, set of alternative answers (referred to as a key in the OU), including a best or preferred answer or answers and some distractors. Often in CMAs the same set of alternatives may be employed for a range of questions. This can be convenient for printing an assignment booklet but the practice often generates difficulties. For example some distractors may be irrelevant in the context of a particular item. Often, the stem of the item is augmented by additional text (for example describing an experiment) which is common to several items. This practice can be both good in that it avoids repetition and bad in that it can lead to successive items becoming linked in such a way as to mean that a wrong answer for one question leads irrevocably to incorrect answers on the succeeding questions.

The UKOU has permitted quite complex question forms employing for example, array formats and rankings. This allows for the questions to be more interesting than the stereotyped true/false questions which are sometimes

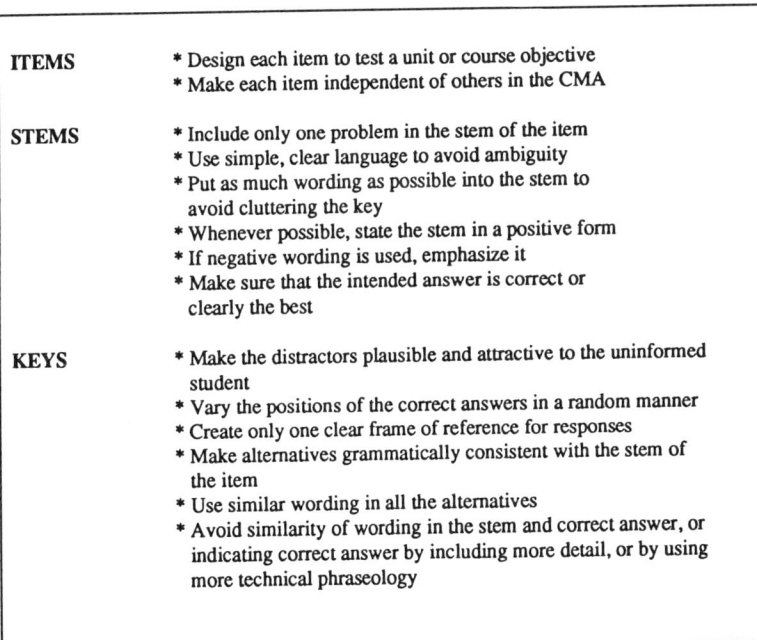

ITEMS	* Design each item to test a unit or course objective * Make each item independent of others in the CMA
STEMS	* Include only one problem in the stem of the item * Use simple, clear language to avoid ambiguity * Put as much wording as possible into the stem to avoid cluttering the key * Whenever possible, state the stem in a positive form * If negative wording is used, emphasize it * Make sure that the intended answer is correct or clearly the best
KEYS	* Make the distractors plausible and attractive to the uninformed student * Vary the positions of the correct answers in a random manner * Create only one clear frame of reference for responses * Make alternatives grammatically consistent with the stem of the item * Use similar wording in all the alternatives * Avoid similarity of wording in the stem and correct answer, or indicating correct answer by including more detail, or by using more technical phraseology

Table 11.1. Rules for constructing simple multiple choice assessment questions (taken from Melton, 1976, Open University internal report).

wrongly assumed to be the only way that multiple choice questions can work. General advice to the question setter of the 'keep it simple' type should not be forgotten however, as it is often the case that complex item formats can confuse students and therefore be of limited effectiveness for assessment purposes. To write sound multiple choice questions is not easy, but there are guidelines which can be used to help produce better questions. Also, a series of reports by the UKOU's Student Assessment Research Group working in the 1980s (Stratton, Armitage and Scott, 1982) described faults they observed in CMAs they analysed, and typified them as syntactic or productive faults.

Examples of rules they coined to avoid syntactic faults are:

- the stem should present a clearly formulated problem and establish an appropriate frame of reference;
- there should be sufficient distractors to act as foils for the best answer;
- the alternatives should be homogeneous (i.e. the form of the correct answer should not look different to the distractors).

Examples of rules to avoid productive faults are:

- do not make the question too easy. An easy question won't discriminate between good and poor students;
- do not allow the item to be affected by prior items;
- make the alternatives mutually exclusive.

These rules were formulated by using the procedures of item analysis, setting performance criteria for item coefficients to select questions which appeared *statistically* to cause students problems. The nature and source of the fault was identified by examining the item itself, the relevant course text and the item coefficients. These item coefficients include the difficulty index (the percentage of students identifying the correct answer to the item or how easy the item is) and the discrimination index (the correlation between student performance on the test item compared to overall performance on the test of which the item forms a part).

As already mentioned formats can be more interesting than the straightforward true/false item. Figure 11.2 shows a table of data which oceanography students are asked to interpret in order to select correct statements from a variety of possibilities. Figure 11.3 shows an array question. Figure 11.4 is an illustration of how a chemistry question which involves a calculation can be extended. Figure 11.5 is an example of one way in which a stratigraphy question in geology can be posed.

Other general points about the use of CMAs include economic arguments, turnaround time, student study time, and reliability. CMAs cost less than tutor-marked assignments (TMAs), CMA turnaround time (the time between a student despatching and receiving feedback) is typically 5 days, as opposed

Q16 Given that the two stations are in the vicinity of a front between two water masses, select two true statements from the key. You may find it helpful to sketch *T* and *S* profiles and/or cross-section(s) between A and B. Pencil across *two* cells in row 16.

TABLE 1

Depth (m)	Station A (1° N)		Station B (3° N)	
	T	*S*	*T*	*S*
5	24	34.0	26	33.0
25	22	34.0	24	33.0
50	16	34.8	17	34.7
100	14	34.9	14.5	34.8
200	13	34.8	14.5	34.8
500	8	34.9	8	34.8

KEY for Q16

A The front slopes down towards the south.

B The characteristics of the front resemble those of the front shown in Figure 4.13 of *Seawater*.

C The density difference between the water masses on either side of the front is due more to the difference in salinity than to the difference in temperature.

D The density difference between the water masses on either side of the front is due more to the difference in temperature than to the difference in salinity.

E To the north of the front E – P is probably negative.

F In this area of the Pacific the permanent thermocline is at a depth of about 200 m.

G In this area of the Pacific the seasonal thermocline is about 25–50 m deep.

H The water at 500 m is largely North Atlantic Deep Water.

Figure 11.2. Question from a third level oceanography course, which requires students to undertake some data analysis (Open University, S330, CMA 01, Q.16, 1992).

ARRAY for Q7–9

	A	B	C	D	E	F	G	H
row 7	Sedimentary rock with fragmental texture	Sedimentary rock with fossils	Coarse-grained igneous rock	Fine-grained igneous rock	Extrusive igneous rock	Volcanic igneous rock	Intrusive igneous rock	Contains gas bubbles
row 8	Has not crystallized from a magma	Slow rate of crystallization from a magma	Fast rate of crystallization from a magma	Low temperature of formation (below 100 °C)	Rock does not contain quartz	Crystallized at Earth's surface	Main rock in ocean crust	Same magma type as basalt
row 9	Main rock in the mantle	Plutonic equivalent of basalt	Plutonic equivalent of rhyolite	Plutonic equivalent of pumice	Volcanic equivalent of gabbro	Rock has no volcanic equivalent	Rock often contains fossils	Rock is metamorphic

Q7–9 This question relates to rock specimen S1, granite, which you examined during the AV sequences in Units 5–6. From the array for Q7–9, select *five* items that are CORRECT for granite. For example, if you choose row 7 columns B and E, row 8 columns B and F and row 9 column D you should pencil across cells B and E in row 7, B and F in row 8, and D in row 9.

Pencil across *five* cells in rows 7–9.

Figure 11.3. Array questions from a foundation level Earth science assignment (Open University, S102, CMA 42, Qs. 7–9, 1991).

to 2–3 weeks for TMAs, and CMAs typically take students much less time to complete.

The main weakness of CMAs is the limitation in what they can test. For example, a CMA cannot test how well students can express themselves, sustain an argument or organize knowledge.

CMAs are very versatile. Given a certain level of skill and imagination on the part of the question writer, we are satisfied that they can be used validly in a wide range of subjects and at all levels of undergraduate work. We acknowledge that there are some purposes for which CMAs are unsuitable. But in many cases the choice between a CMA and a TMA often cannot be made in terms of purposes or the type of knowledge to be assessed. In such cases, whether a sound assessment of a student's knowledge is made is much more dependent on how well the assignment is set and marked than whether it is a CMA or a TMA.

(Open University, 1978)

Tutor-marked assessment

Under this category, we consider the regular written assignments that students are required to submit to their tutor. In the UKOU, TMAs can take a number of different forms. In arts, social sciences and education subjects, they usually consist of an essay question requiring an answer of around 1500 words. In mathematics, science and technology assignments, TMAs are more varied but typically might contain three questions; an individual question might require a short essay answer, a description, a calculation, a problem solution or a proof. Figure 11.6 shows some examples. TMAs are marked by the same tutors who provide face-to-face tutorials. This gives the University a large task in ensuring that the marks awarded are reliable, i.e. that there is not too much variation

---

PART D

The questions in this part relate to Unit 16 and carry 25% of the marks for this assignment.

Q10 and Q11 An oxygen atom (O) reacts with a molecule of ozone (O_3), in a reaction that occurs naturally in the ozone layer; Figure 1 is a reaction coordinate diagram for this reaction.

$$O(g) + O_3(g) \rightleftharpoons O_2(g) + O_2(g) \qquad (4)$$

Q10 What is the value for the enthalpy change for reaction 4? Select *two* CORRECT items from the key for Q10 and Q11, *either* A or B *and* one of the options C to G.

Pencil across *two* cells in row 10.

KEY for Q10 and Q11

A	Positive	E	$370\,kJ\,mol^{-1}$
B	Negative	F	$390\,kJ\,mol^{-1}$
C	$20\,kJ\,mol^{-1}$	G	$410\,kJ\,mol^{-1}$
D	$40\,kJ\,mol^{-1}$		

Q11 What is the activation energy for the forward reaction 4? Select *two* CORRECT items from the key for Q10 and Q11, *either* A or B *and* one of the options C to G.

Pencil across *two* cells in row 11.

Q12 What would be the effect of an efficient catalyst on the value of ΔH for reaction 4? Select the *one* CORRECT item from the key for Q12.

KEY for Q12

A It would have no effect on ΔH.

B It would increase ΔH.

C It would decrease ΔH.

D It would cause ΔH to change sign.

E It is impossible to decide from the information given.

Pencil across *one* cell in row 12.

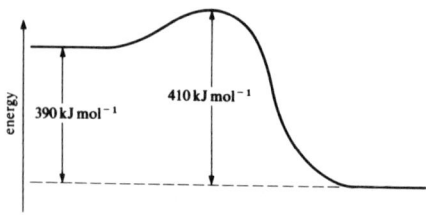

FIGURE 1 Reaction-coordinate diagram for reaction 4 (not to scale).

Figure 11.4. Set of questions from a foundation level chemistry assignment (Open University, S101, CMA 45, Qs. 10–12, 1989).

Q10 What is the *youngest* geological event represented in the cross-section? Select *one* CORRECT answer from the key for Q10.

KEY for Q10

A   Formation of igneous body W

B   Folding of the mudrocks, siltstones and **sandstones** with banded ironstones

C   Formation of igneous body Y

D   Deposition of the limestone with abundant **echinoids**

E   Formation of igneous body X

F   Faulting F–F′

G   Erosion of the folded mudrocks, siltstones and sandstones with banded ironstones

H   Tilting of the rock sequence which contains the mudrock with abundant plant remains

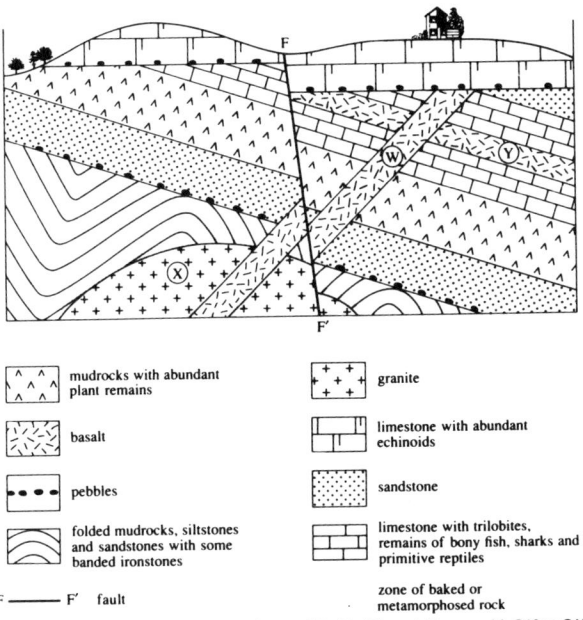

mudrocks with abundant plant remains

basalt

pebbles

folded mudrocks, siltstones and sandstones with some banded ironstones

F ———— F′   fault

granite

limestone with abundant echinoids

sandstone

limestone with trilobites, remains of bony fish, sharks and primitive reptiles

zone of baked or metamorphosed rock

FIGURE 2   Sketch cross-section through part of the Earth's crust. For use with Q10 to Q13.

Figure 11.5. Question from a foundation level Earth science assignment, based on a cross-sectional diagram (Open University, S102, CMA 49, Q. 10, 1990).

in marks awarded by different tutors. When double marking experiments have been conducted (Edwards, 1979), assignments in the mathematics, science and technology style have shown a reasonable degree of consistency in grading, though long essay questions are more problematic in this respect.

The tutors will have to mark the students' attempts at the questions by following some marking guidelines produced by the TMA author. This means that any room for interpretation in the marking notes can cause lots of problems. There are three sources of unreliability of assessment information:

Question 1

This question is related to Unit 13 and carries 30% of the marks for this assignment.

In 1937, the English physicist Paul Dirac suggested that the value of Planck's 'constant' h may not always have had its currently accepted value of 6.63×10^{-34} J s. The idea was that the value of this quantity may have changed significantly as the Universe evolved. Although there is as yet no clear empirical evidence to support this hypothesis, it is amusing to contemplate the experimental results that would be observed in regions of the Universe in which h has a value that is significantly different from the currently accepted value on Earth.

Imagine a galaxy in which the theories that have been formulated on Earth are correct, in which all the fundamental constants (ε_0, μ_0, c, k, e, m_e, ..., etc.) are the same as ours on Earth, but in which the value of their Planck constant (h', say) is *double* that of ours (h). Suppose that you are in contact with a group of alien scientists in this galaxy who have a well-equipped laboratory, who use units of mass, length, time, etc., identical to ours, who speak perfect English, but who know nothing about quantum mechanics. In order to help them to begin research into this fascinating subject, you advise them to investigate the photoelectric effect in the way described in the Text of Unit 13.

(a) Discuss briefly the way (or ways) in which the aliens' results concerning the photoelectric effect would *differ* from those obtained on Earth. Also, discuss briefly the way (or ways) in which their results would be *the same* as those obtained on Earth. *(12 marks)*

(b) Describe very briefly with the aid of a sketch-graph how the aliens could use their data on the photoelectric effect to measure their value of h'. *(6 marks)*

(c) Some of the observable properties of the hydrogen atoms in the aliens' laboratory will differ from the observable properties of hydrogen atoms on Earth. State *one* such observable property, and state exactly by how much the *aliens'* determination of the property should differ from a determination of the same property made by an observer on *Earth*. *(6 marks)*

(d) Do you expect that the following experiments will give the *same* results in the aliens' laboratory as an analogous experiment carried out on Earth? For each experiment, give (in one sentence) a reason for your answer.

(i) An experiment to measure the wavelengths of the characteristic X-rays of tungsten.

(ii) An experiment to measure the electrostatic force between two electrically charged metal spheres. *(6 marks)*

Figure 11.6. Question from a tutor-marked assignment for a second level physics course (Open University, S271, TMA 04, Q.1, 1992).

---

poor question design, variable conditions under which students can complete the assessment and marker variation. The last issue is one that can be addressed by the provision of helpful marking notes and arrangements for some kind of monitoring or double marking. It is also very worthwhile to make some investment in tutor briefing or training to remove different interpretations on the part of the tutors about the way the marking function is to be performed. There is strong empirical evidence to suggest that every additional marker, even in science subjects, increases the reliability of assessment, although e.g. Pillener (1969) has questioned the precise way that extra markers might affect scores. Studies have shown that many factors can influence the grades that markers award: handwriting, general standard of presentation and verbal fluency can significantly affect a student's grade though these in fact may be irrelevant to what was being tested. The role of the tutor in marking students' continuous assessment work will be examined in more detail in Chapter 13.

Practical work and projects

One particular use of TMAs in science is that they provide an opportunity for students to have feedback on the way in which they carried out their experimental work and interpreted the results.

Sometimes students have to complete more extended pieces of work, collecting and analysing original data or writing an extended essay. The TMA and examination system can be adapted to provide a scheme suitable for the assessment of projects. Edwards, in a 1982 study of the reliability of project marking, assesses it to be at about the same order of magnitude as for essay questions in conventional assignments. But while admitting that the reliability of grades is a problem, Edwards asserts

> Though short questions and multiple choice questions produce more reliable marks, they do not test higher-order skills that the more open assignments do. Projects, in particular, force students to use their initiative and develop enquiry skills.
>
> (Edwards, 1982, p. 33)

One successful approach to the assessment of project work adopted in the UKOU science Faculty is the 'double TMA'. For example in a third level course on *Ecology* (see Ch. 10) students are allocated to specialist tutors on the basis of a brief outline of what project they intend to do. They are then required to write a detailed plan of anticipated problems and how they propose to solve them. After making initial contact with the tutor, the student carries out preliminary work, consulting the tutor if necessary and then writes a draft report of a maximum of 2,000 words to which the tutor awards a grade. This tutor also makes suggestions for additional practical work, changes to presentation, etc. The student completes the project and submits a final report to a second independent tutor, who also awards a grade. Evaluation of this scheme suggests that it works well, and shows that the marks from each tutor differed by less than one grade for most students.

Monitoring assessment

CMAs are relatively easy to monitor and evaluate, since the computer record itself can be used to set up a system of item analysis. This can identify questions that have performed well in terms of assessing students' understanding and have discriminated between the responses of stronger and weaker students. Good questions can be conserved for future use. Bad questions can be revised or discarded.

Procedures for monitoring tutors' marking, in respect of both the grade awarded and the feedback provided for students, are more cumbersome, but must be put in place in a distance education institution to ensure consistency. Methods of doing this will be discussed further in Chapter 13, in relation to the role of the tutor.

Validation of examination standards is of course particularly crucial. In the British system (including the UKOU) this is carried out by an external examiner (i.e. an academic with a reputation in the area of the course who is from a different institution) who approves examination papers and reviews marked scripts. The large population courses, with hundreds or thousands of students, such as occur in distance education institutions like the UKOU, present special problems for external examiners. It is, for example, impossible for them to see more than a small proportion of the students' scripts. In the UKOU, the computer system is used to collate results and present to external examiners sufficient information to allow them to construct the pass list in an informed way. The UKOU was among the first British institutions (Kirkwood, 1992) to develop a simple appeals procedure for both examinations and continuous assessment, which examines any written request from the student for a reconsideration of the results.

6. CONCLUSIONS

It is difficult to overestimate the influence of assessment practices on students' learning. Particularly in distance learning, students use assignments as a source of information about what to concentrate on and spend time on, and the feedback on their performance to help guide their future study. This influence is apparent in the reaction of one remote student to receiving the grade for her first assignment:

> I was dreading my first assignment. I thought all I wanted was a pass. The first grade I got was a credit. I was jumping up and down with glee. I mean a credit is very ordinary. We know that now, but in those days to have got a credit with my essay was tremendous. The fact that I had done it at all, I could have had a failure, couldn't I? If I did fail, I would not have gone on because even though I was determined, I also couldn't have coped with rejection.
>
> (Evans, 1993, p. 60)

As a result, enormous effort is expended in the design and operation of assessment systems. However, it is important to bear in mind that all test instruments have weaknesses. 'Tests are not mysterious sources of information, miraculously perfect and fair' (ETS, 1991, p. 3). Tests are simply a source of information about what students were able to do and willing to try to do on a particular occasion, in relation to what the test was designed to measure. It is important that all who engage in teaching and learning develop a realistic perspective on what the variety of grades they award and receive really mean.

It almost is unnecessary to say that the most important feature of an assessment system is that the learners should feel that the procedures used to grade them are fair. Second in importance is that the learners should feel that the work they are asked to do for assessment purposes has some worth in itself, and is of benefit to them as part of the learning process:

> I could easily have given up, a couple of times with the OU, and not sent

in assignments. But the benefit I gained from sending in every assignment, not skipping any, was so valuable.

(Lunneborg, 1993, p. 55)

As noted in this chapter, some things are particularly difficult to achieve when assessing science learning. One key issue in science-based assessment is the best way in which to measure students' abilities in relation to the processes of science (see e.g. Fairbrother, 1988). Another is the discrepancy between students' understanding of science concepts and the extent to which this can be probed during formative assessment procedures designed to uncover where learners are experiencing problems (see e.g. Murphy, 1988).

Another interesting issue related to assessment in general is the degree to which the use of self-assessment can play a constructive part in developing new approaches to assessment. One method used in the decision sciences department at the Georgia State University Atlanta involved students in composing test questions in a multiple choice format; the instructor then created an examination from the class questions. The students marked all the incorrect answers they could identify for each question. The test was validated through group discussion, culminating in a consensus answer key (correct answer). Tests showed significant learning gains and a strong student preference for this methodology (Lewis and Faziollahi, 1989).

Another new approach is the 'records of achievement' or 'profiling' approach as described in Assiter and Swann (1993). This is part of a general move away from norm referenced testing towards a system of identifying and describing the skills that students already have.

Involving students and employees in monitoring, reviewing and recording their own learning and experience is consistent with the growing emphasis being placed by both educators and employers on the development of autonomy in learning.

(Assiter and Swann, 1993, p. 15)

Another reason for the interest in this approach is the increasing emphasis in higher education on skills and capabilities (e.g. the UK national movements such as Higher Education for Capability, and the Enterprise in Higher Education). McCulloch discusses integrating the records of achievement approach into a Higher National Diploma course in chemistry and stresses the importance of the reciprocity element in records of achievement – that is the need for tutors and students to share expectations of each other. This account also highlights a way to approach the problematic area of assessment on projects which involve collaborative working. Raggett (1994) considers that the outcomes model will become more important in higher education. This model asserts

that successful participants, employers and other interested parties should know what a qualification or a successful learning experience means in terms of what the holder should know, understand and be able to do.

Outcomes provide the criteria against which learners are assessed, but they make no prescriptions about the modes of learning or where it should take place. It opens the way for learners to present evidence of successful learning however or wherever it has been achieved.

(Raggett, 1994, p. 139)

An example of the learning outcomes model is given by the National/Scottish Vocational Qualifications, which at higher levels are increasingly being offered in UK colleges and universities. A current focus of interest in this area is on ways of assessing other sorts of outcomes – for example, affective rather than cognitive outcomes.

There is no question that assessment is a key feature of the experience of learning in open and distance learning. In this chapter we have highlighted the range of functions that it can serve: both formative or diagnostic (providing feedback, pacing and motivation for the learner) and summative (contributing to the grading process). But as assessment plays such a vital part in students' learning experience, the following words of caution are worth taking to heart:

It is probably safe to say that no university in the world has a satisfactory assessment system. There may of course be thousands of academics who are sanguine about the assessment procedures that they use, and who cannot see any way in which their procedures might be improved. But it is a commonplace observation that the results of formal assessment procedures rarely reflect either the absolute merit or the relative merit of the students involved. In addition, there are good reasons for believing that – irrespective of whether or not the student population ends up ranked in an equitable way – formal assessment procedures can fail to reveal to the universities concerned the – possibly – crass inadequacies of their teaching.

(Lewis, 1972, p. 109)

12

PUTTING IT ALL TOGETHER: THE COURSE TEAM APPROACH

The majority of course team members are anxious to see the task through and . . . they draft their units, respond to criticism, redraft them and eventually hand them over to publishing (occasionally, would you believe it, on time). The miracle is that this happens in course after course, year after year.

(Blowers, 1979, p. 55)

1. INTRODUCTION

The very complexity of a multiple media distance education course virtually dictates, as discussed briefly in Chapter 4, that it cannot be produced by a lone individual, but must involve a number of people, with different specialisms. If the various course components are to be well integrated, the content arranged logically and the learners' workload distributed fairly evenly throughout the course, then those people cannot generate the separate course components in isolation, but have to function as a team. In this chapter, we consider the processes of whole course production, and some of the effects that group dynamics within the team may have on these processes.

The nature and composition of the team may vary according to circumstance. The *large team*, as pioneered by the UKOU, includes academics, who are the subject matter experts, media specialists, educational technologists, editors and a 'course manager', who services the team and organizes the administrative aspects of the production. This kind of team is normally chaired by one of the academics, although the team as a whole operates on a fairly democratic basis, within institution-imposed restraints of timescale and cost. The editors liaise with graphic designers, illustrators and printers. The media specialists liaise with camera crews, sound technicians, software designers, etc. All members of the team are expected to read and comment on one another's material, and to take responsibility for the course as a whole. Within this broad framework, many variations are possible. The academic experts may include both 'writing' and 'reading' members. Consultants, either external or internal to the institution, may be brought in to contribute to specific parts of the package or particular stages in the planning/production cycle. By and large, the UKOU has stuck to this model throughout its first quarter century, and the concept has gained international prominence as a result. The large team has many strengths. Because of the breadth of expertise of its members and the number of quality checks it imposes, it can aim at producing very high quality materials, in terms of both content and presentation. The momentum of the team as a whole carries along its individual members towards the completion of their components of the task. However, the large team also has weakenesses: it can be unwieldy and slow to react to new demands, it can splinter into factions or become prone to 'bandwagonning', or be effectively hamstrung by a few recalcitrant individuals. We will examine some of the advantages and problems in more detail in Section 3.

Other distance education institutions have successfully adopted rather different approaches, summaries of which have been given by Kaye and Rumble (1981), Dodds (1993) and Hawkridge (1994). Closest to the standard UKOU model, and evolved from it, is the *lean team* adopted at Athabasca University, in which each type of role is covered by a single person. Thus a typical lean team might consist of one academic or content consultant, one instructional designer, one editor, one graphics designer, and one media consultant. A very significant feature of this kind of team is the higher profile it gives to the educational technologist, who indeed often assumes the managerial mantle.

> In an academic-heavy team, the instructional developer is a lone voice and may well be little more than a token participant. In the Athabasca team, however, the developer is generally a very active participant, and the team operates on the basis of an equal representation of all functions so that in terms of weight of numbers no sector can over-ride another.
>
> (Stringer, 1980, p. 14)

The steamlining of course production along these lines has obvious advantages in terms of flexibility and cost effectiveness. Stringer, however, also lists several disadvantages, chief among which are the dangers of leaving virtually the whole responsibility for academic content and accuracy with a single academic, and the lack of latitude or of a suitable forum in which to explore alternative instructional strategies.

The *transformer team* model (Melton, 1990) divides the course production into two stages: an initiation phase, in which subject matter specialists prepare drafts that define the academic content of the course, and a transforming phase, in which educational technologists work in conjunction with media experts and editors to convert the material into a complete multiple media package suitable for distance learning. Variations on this model may also be useful when external consultants are commissioned to write material, as however carefully they are briefed they may not conform in all respects to the 'house style' of the institution, and in any case will probably not have the opportunity to see the rest of the course material into which their contribution must fit. Under such circumstances, it is often better to commission a 'first draft only' from them, leaving a core team from the institution to rewrite as appropriate in order to ensure pedagogic continuity across the course. *Joint teams*, involving collaboration across several institutions, are another variant on the transformer team model, in which core material produced in conjuction may then be adapted in a variety of ways to suit the different needs of the participating institutions. This approach offers particularly interesting possibilities for collaborative ventures involving both the distance teaching and mainstream sectors.

The *wrap-around team* may be appropriate when the course can be based on an existing item, such as a broadcast television series or a conventional textbook. At its most basic, the team simply produces study guides to make the main item more accessible to remote learners, and assignments so that the course can be studied for credit. More ambitious projects can add other new elements – additional text, cassettes, tutorial provision, or whatever – in which case the team may take on some of the characteristics of the standard (large) course team. Wrap-around teams based on broadcast TV have been quite successful in the USA, linking community college programmes with educational TV stations, but similar ventures in the UK (Freeman, 1979) have experienced more problems. Wrap-around teams building courses on conventional textbooks are fairly popular, and the UKOU has used this mode of production to generate relatively low resource courses for small student populations – one example being the third level course on electromagnetism described in Chapter 6.

2. COURSE DESIGN AND PRODUCTION

In this section and the next, we will illustrate a typical methodology of course production by a 'standard' course team. To examplify the various stages, we will mostly use case study material from a particular UKOU course, *Discovering Physics*. Reference to this particular course has already been made in several previous chapters and Figures 5.5 and 5.7 are also relevant here. In Chapter 14, we will describe some of the evaluation studies carried out on this same course.

2.1. Initial planning stages

Even before the course team is formed, significant decisions are likely to have been made at the institutional level about subject and length of the course, its overall aims and place within the profile, and the production schedule. Other questions that have to be answered, at least in broad terms, early in the planning stages include:

- What will be the students' entry behaviour? What knowledge and skills can be assumed? Will *prerequisite* courses be specified, and if they are, is it safe or fair to assume that the students will have full mastery of all the material in those courses?
- What is the *content*, and at what level will it be presented? Will the course itself be a prerequisite for other established higher level courses, so fixing some of the content as essential?
- What steps can be taken to ensure that the content does not become *dated* during the expected lifetime of the course?
- What are the main *objectives* that the students will be expected to achieve?
- Can the course be based on or wrapped around an *existing text* (or other item)?
- How will be material be *sequenced*? To what extent does the content dictate the hierarchy, and to what extent can the ordering be flexible?
- What *teaching strategies* will be used and what will be the function of individual *media* within these strategies?
- What *assessment* of learners' progress will be made?

Prerequisites

The question 'to what cohort of students will the course be addressed?' is obviously a crucial one with which to start. The distinction between concepts or skills that will be assumed and those that will be developed as an integral part of the course has to be established from the outset of the planning process. If the course population is to be drawn from a wide variety of different backgrounds, or is to serve as an introduction to students with no previous experience of the subject, then some preparatory material may have to be included in the package. To help students make the most effective use of such preparatory material, a (self-administered) diagnostic pre-test may be a useful device.

Content

It often seems that course teams designing and writing courses spend most of their discussion time on issues of content. However, a more detailed analysis shows that in fact the broad categories of conceptual content are usually decided fairly quickly, especially for courses covering traditional areas for which the main decisions to be made may simply centre around how far to go in the conceptual development. A lot of the debates tend to be really either about the fine structure of the teaching and in particular the sequencing of material, or about skills levels, or about the choice of examples to illustrate the concepts. The illustrative material is of course extremely important, as it is this that provides the 'storyline' for audiovisual or multimedia components, and it plays a crucial role in maintaining students' interest and commitment.

One type of course that does require detailed discussion of content, however, is the issues-based course, i.e. the kind of course which is built entirely around case studies, and in which all the conceptual and skills development is carried out in the context of those cases. Again, the instructors' views of what will interest and motivate students play a large part in the design choices. The interactions within the team are also crucially important, as people who are only interested in their own specialism can be a particular handicap on interdisciplinary courses and on courses that cross traditional subject boundaries, and may bring about conflicts over content. In such situations, the imposition of some sort of corporate will can sometimes be the only means of setting the course aims and strategy.

Objectives and assessment
Riley (1984b) noticed a surprising lack of fit between the traditional methodology of instructional design and the actual practice in many course teams. It was her observation that the writing of (student) objectives and assessment material did not feature in the early planning stages. (Indeed, it is our experience of UKOU course teams that the preparation of assessment material is a task undertaken in conjunction with, or even after, the writing of the *final* drafts of teaching material rather than the initial ones.)

Dating
High production costs (whether in terms of writing time or media development) usually means that courses are expected to have extended lifetimes – typically of the order of a decade at the UKOU for example. This prospect can lead to course designers sticking to standard material, through fear that new or controversial approaches might turn out not to stand the test of time. While many subject areas within science are indeed unlikely to change in the forseeable future, limiting course material entirely to such topics would be stultifying for both instructors and learners, and give students a quite erroneous impression of science as a base of unchallengeable truths.

One possible way round these difficulties is the inclusion within the course package of a 'yearbook' or similar item, which can be updated annually (or as often as required) to incorporate new and topical material, or to introduce issues of current debate. An alternative type of option is to bind the printed text into numerous small booklets rather than one large one, to send it out in loose leaf format ready perforated for insertion into ring binders, or even to design the course as a series of free-standing modules, so making it easier to rewrite small portions of the course from one presentation to the next. However, these kinds of strategy also carry the disadvantage that the individual parts of the text have to be fairly well decoupled from one another, reducing the possibilities for integration between the various elements of the course as a whole and making it more difficult to foster a smooth development of skills.

Sequencing
The role of macro-strategies in instruction is not well understood.
(Van Patten, Chao and Reigeluth, 1986, p. 466)

Rowntree (1990) has listed a number of sequencing structures, and these have also been discussed by Lawless (1994). Figure 12.1 is based on their classification schemes. In any extended piece of teaching, it is likely that a variety of such devices will be used, but in most cases there is no single 'right' order although it is only too easy for instructors to assume that there is a 'natural' sequence in which particular types of content should be presented. While it is true that many concepts and skills in science are indeed hierarchical, careful reflection may reveal that an apparently 'obvious' order is no more than the 'conventional' order, or the order in which the instructors were themselves taught the topics, and not a sequence necessarily dictated by the content itself. Once this possibility is recognized, it may open the way for more imaginative treatments of 'standard' topics – which might for example allow students a more personal route into the material through their previous ideas and experiences, or build up motivation and enthusiasm by introducing 'frontiers of knowledge' science at the beginning of the course rather than at the end.

- **Topic-by-topic** sequencing occurs in modular courses, in which modules may be studied in any order. (Within each module, a different type of sequencing would probably be appropriate.) The difficulty for the instructor lies in choice of topics to form a coherent whole and in enabling students to understand the links and relationships between topics. (See also problem-centered sequencing.)

- **Place to place** sequencing starts with one 'place' and moves to an adjacent one (e.g. in a description of the geological features of an area), although the 'places' need not be geographical locations: the method may be useful in describing anything in which spatial relationships are important. In **concentric circles** sequencing, each 'place' is included within the next.

- **Chronological order** is an appropriate way to present the historical development of scientific discoveries, or to describe the stages in a process, but any extended adherence to such sequencing may inhibit the proper exploration of underlying concepts.

- **Structural logic** dictates the sequencing of hierarchical subject matter. **Causal sequencing** is a variant on this, and is also related to chronological ordering.

- **Problem-centered** sequencing start by presenting students with a problem, the exploration or solution of which provides a context in which to develop concepts and skills. This kind of case study approach is particularly appropriate for issues based courses.

- **Spiral sequencing** introduces a number of concepts and their relationships, and revisits them all several times, with increasing complexity or sophistication of approach at each successive stage.

- **Backward chaining** is a sequence by which the final step in a chain is introduced first, and a proccess taught 'in reverse'.

Figure 12.1 Sequence models (based on analysis by Rowntree (1990) and Lawless (1994)).

Course designers also need to be aware that the order in which they decide to present material to learners, and the order in which the learners actually choose to study it may differ considerably. Research by Mager (1961) into undergraduates' approaches to the study of electronics showed that they rarely built their understanding from the parts to the whole in the way most often followed by instructors. Instead, they started from examples of wholes that interested them and continually attempted to relate new information and concepts to their previous knowledge. This finding supports Ausubel's model of cognitive psychology and the constructivist view of learning discussed in Chapter 3, in suggesting that the most important factor in learning is the student's prior conceptual understanding. Taking full account of this may result in a sequencing very different to that which might be laid out by a subject specialist.

A different kind of problem may arise in multidiscipline, or issues-based courses, where

> content is frequently chosen and assembled specifically for an individual course and its development and the clarification of the interrelationships is part of the process of constructing the course. This can be an intellectual adventure for the course designers – combining ideas and concepts in novel patterns and planning to present them in innovative ways. While this may be an interesting experience for the course designers, the danger is that . . . the innovative, even tentative nature of the course structure may not be at all apparent to students who may find it more inaccessible to the extent that it was exciting for the course designers to make.
>
> (Lawless, 1994, p. 58)

2.2. Media selection

We have already discussed the appropriateness of different technologies for different applications – what might be called 'matching the medium to the message' – for example, the use of video for effective demonstration of phenomena based on time-lapse photography, or for providing a vicarious experience of field work in a remote and hostile environment. On the other hand, there is also evidence, as discussed in Chapter 4, that should lead course designers to beware of a presumption of superiority for any particular method of delivery, and to evaluate carefully the learning outcomes of each piece of teaching separately from the medium in which it is embedded. However, there are certainly good arguments for a media *mix* being a powerful aid to learning, as total reliance on a single method of delivery can bring its own problems. For example, there is the sense in which

> the printed word alone inevitably imposes its own peculiar *imprimatur* of validity and truth . . . it is a curious fact about the written

word that humans (professional scientists and laymen alike) seem to find it hard to extract anything but black and white conclusions from it, no matter how many caveats and conditional clauses are included.

(Dunbar, 1982, pp. 68 and 70)

It is therefore desirable that textual presentations be balanced by media with a more conversational flavour. Audiovision, for instance, can inject a certain amount of informality; if it is well presented the student will get some sense of personal contact with a course author, and the medium is well suited to the kind of conversational aside that can enliven teaching or learning but for which it is difficult to set just the right tone in print.

One of the most obvious difficulties for those constructing distance education material is the lack of feedback from students, coupled to the fact that once the course is produced changes are likely to be slow or costly or both. As Rowntree (1990, p. 161) has put it, 'being a distance teacher is never being able to say: "I'm sorry – let me try to express that another way"'. At least partly for this reason, some redundancy of material – in the sense of a repetition of ideas in different media – is often beneficial. This allows students who have trouble in, say, following an oral argument to see its main points laid out in a written summary. There is also much to commend in what is sometimes called the 'two bites of the cherry' principle – i.e. the reinforcement of learning that can come from the same concept being approached several times in different ways and with different emphasis.

Often the media mix (and/or the proportions of its ingredients) is dictated mainly by pragmatic, rather than pedagogic, considerations; cultural and economic factors will also have a bearing on media selection. In some cases institutional practices may impose particular media mixes; for example if a course is allocated a certain amount of TV air time from the outset, then the choices to be made essentially boil down to exploiting the medium of TV to best advantage in pursuit of the course aims. If on the other hand, TV production resources have to be secured in a competitive bidding process, or traded against resources for other components such as a CD-ROM or HEK, then the various strengths and weaknesses of the alternative media have to be examined in relation to the concepts and skills that the course sets out to develop. There may also be a limit to the number of different media that a student can reasonably juggle in working through the package.

The most important thing is that the various media should be integrated into a coherent package, although this does not preclude (especially at higher levels), a structure that actively requires students to undertake additional reading, or to synthesize material for themselves. The structure of the *Discovering Physics* course is outlined in Figure 12.2. The Main Texts carry virtually all the study advice, the references out to other components and the exercises associated with them. (For example, Figure 7.3 illustrates how one of the video sequences is integrated with the text.) Additional components not shown in Figure 12.2

included three 'summary cards', which listed the most important equations for each block and could be used as ready reference bookmarks, and a Course Summary and Revision Guide, which was interfaced with an audiosequence designed to help students prepare for the final examination.

2.3. Course outlines and specifications

After the initial round of discussions, the content and sequencing will have been agreed. Sections or tasks can be allocated to team members, culminating in the writing of outlines for each part of the course, which set out the intended content and teaching strategy of each component. These outlines are best discussed all together if possible, rather than on a series of different occasions, as there will be inevitable mismatches between the material produced by different authors which will need to be reconciled. After that, it may be possible to produce reasonably detailed specifications concerning, for example,

- conventions to be adopted as regards access devices;
- interfacing of the various components;
- notation;
- layout and typographically signalled structures.

We have already discussed access structures in Chapter 6; for example Figure

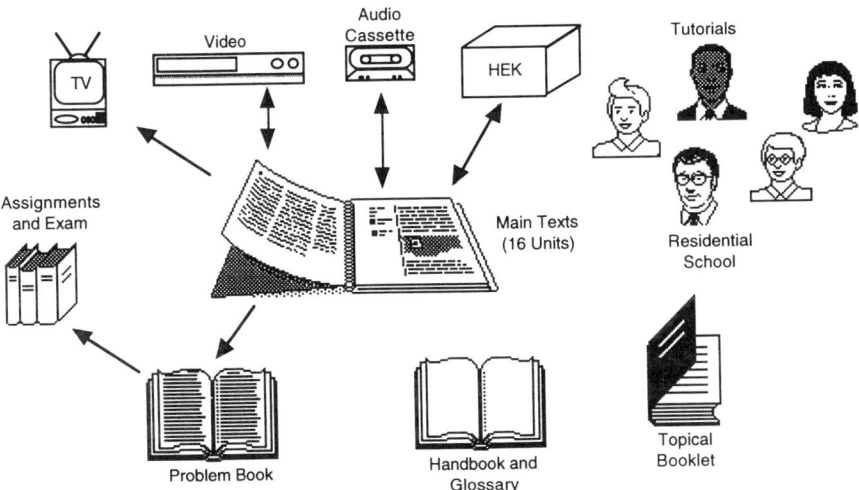

Figure 12.2. Structure of the UKOU course *Discovering Physics*. Summaries of TV programmes are included in the Main Texts, as are instructions for home experiments. Use of the videocassette involves students in carrying out exercises detailed in the text in between watching videosequences.

Study Guide	This should identify the major components of the Unit (TV/video, AV, Home Experiment). If students need to gather bits and pieces for an experiment, they should be warned about this. Included as part of the Study Guide should be a (marginal) Table 1, which gives a suggested division of study time among sections in terms of percentages. If authors wish to offer advice about possible short cuts, this should appear in the Study Guide.
Section summaries	Virtually all sections should have a summary as a numbered subsection (with the main exceptions being introductory sections) Summaries should cover both content and process/skills. Summaries may consist of short numbered points, or be written in a more discursive style, but either way should make it clear to students what they are expected to be able to <u>do</u> when they have completed a section. End-of-section SAQs should follow the summary (see also notes on SAQs)
ITQs	ITQs are an <u>essential</u> part of students' work on the Unit. These are the questions that they must do as they come to them, in order to make sense of the next piece of text. By definition, ITQs will normally occur in the middle of (sub)sections [cf. SAQs below].
SAQs	Given the additional devices of Worked Examples and Problems, SAQs should not be too complex, and should certainly include some quick, simple, even qualitative, questions designed to test students' grasp of the <u>basic</u> principles. The more difficult SAQs (a few per Unit) should be starred. SAQs will normally occur only at the end of subsections and sections. The former type will test the material in that particular subsection. End-of-section SAQs, testing material from anywhere in the section, should be placed after the section summary.
Objectives	Each objective should be keyed to at least one ITQ or SAQ or Worked Example or Problem. (This may lead to slight artificiality, but will ensure that all objectives are testable and covered with the Unit!) One question may cover more than one objective.

Figure 12.3. Part of the checklist drawn up by *Discovering Physics* course team, covering the way access structures, such as 'ITQs' (in-text questions) and 'SAQs' (self-assessment questions), were to be incorporated into the Main Texts.

6.3 illustrates the Study Guide for one of the units of the *Discovering Physics* course, and it was agreed among the course authors that every unit would begin with orientation advice laid out in a similar style. Apart from making the course look more of a coherent whole, and easing the task of the editor, there are definite advantages for students in having a consistent layout of study comments, self-assessment questions and other access devices. An extract from the guidelines for authors on this particular course is shown in Figure 12.3.

Similar sets of guidelines can be devised for the formal structures by which the various media are interfaced with one another, as exemplified in Figure 12.4.

Notational styles (e.g. for symbolic representation of physical quantities) need to be fixed as early as possible in the drafting process, to ensure consistency, and to avoid excessive amounts of re-keying at later stages. If the documents are being generated and eventually published electronically, the editor can set up electronic style sheets for the layout of all typographic devices. There are also pedagogically based decisions to be made about some aspects of the layout, for example where and in what format answers to in-text questions should be supplied to students (see Stoane and Stoane (1981), for a discussion of alternative methods and their merits and drawbacks in distance education).

Many of these specifications may be refined or even radically changed once the first few units have appeared in extended draft, and the team is better able to see how its initial ideas work out in practice. However, the fact that a few alterations may be necessary later in no way diminishes the importance of the preparation stage. If the course strategy is not clearly defined, and if authors do not have a reasonable feel for the structure as well as the content of the parts that others will be writing, then the first full drafts will not mesh into a cohesive whole. Protracted debates can then become necessary, and these can become acrimonious, especially if authors' investment of time and effort in what they have already produced leads them to resist further changes. Tensions may also be exacerbated by ever-approaching deadlines. One way

Video notes	These will be in the body of the text, at the most appropriate place and labelled as a separate section or subsection. Material covered in the video should not be repeated at length in the text, though a short aide-mémoire may be appropriate. About 30 minutes study time has been allocated to the accompanying exercises, in addition to the 15 minutes actually required to watch the video once through. The Video Steering Group recommends that questions be posed in a very general way on the video itself, followed by a screen graphic/sound effect to indicate a tape stop. The questions should then be set out in detail in the text and given ITQ status (in accordance with the Course Team policy of labelling as ITQs those questions that must be tackled at the point at which they appear in the text). Video-based exercises should therefore be numbered in sequence with the other ITQs in the Unit.

Figure 12.4. Part of the guidelines for authors on the *Discovering Physics* course, drawn up by a subgroup of the course team (the Video Steering Group) and covering the interfacing of the video strand with the Main Texts (See also Figure 7.3).

to cut through these difficulties is to adopt some variant on the 'transformer team' model, involving a change of authorship after the production of the first drafts (see e.g. Monk and O'Shea, 1981), but because this strategy reduces the sense of 'ownership' that authors feel for their material it may be imposed at the expense of commitment.

Drake has further suggested that although the (large) course team can generate an exciting environment for academic discussion, this can in itself lead to problems: the planning stages for the course may be so exhilarating that it becomes increasingly hard to get down to the grind of actually writing the material, and the production falls behind schedule. The 'many hands make light work' approach – increasing the number of academics on the team so as to give each a smaller writing load – may actually exacerbate the problem by prolonging the planning phase, since as Drake (1979) points out (p. 51) 'more minds means more ideas, more possibilities, more futures to discuss'. The other approach – simply giving the team more time to complete its work – adds to the danger of the course being dated almost before it comes out, and also poses difficulties for individuals who are unable to sustain a commitment to the team for the necessary period of time (often years).

3. COURSE PRODUCTION: FROM OUTLINE TO HANDOVER

Drafting stages

Once the broad content and teaching strategy are agreed, authors have to embark on the preparation of their first extended drafts. Most materials for UKOU courses go through at least two, often three, extended drafts before being handed over to the editor and the design professionals. Although the schedules may dictate the exact nature of the drafts (for example if a TV programme has to be filmed at a particular season of the year and its storyline integrated with the text), most course teams draw up for themselves the details of what components they expect to be incorporated in each successive draft. The guidelines agreed by the *Discovering Physics* course team are shown in Figure 12.5. Copies of the main text of each unit and all associated supplementary material are usually circulated to every course team member at each stage, and discussed at a meeting, with written comments also being passed on to the author.

The nature of authorship in a team setting

The course team provides a forum in which much collective creative effort may go into planning, but when the course finally comes to be written, each author is still on his or her own, facing a blank sheet of paper or word processor screen. Stanford (1980) has speculated that while academic

Guidelines for authors

First drafts should contain:
  (i)   First full draft of main text, including:
        study guide, worked examples in two column format, ITQs/SAQs and answers,
        problem(s) and answers, roughs of all diagrams, section summaries;
  (ii) First draft of Home Experiment Notes;
  (iii) Outline of AV content, with first roughs for AV frames;
  (iv) Outline of TV/video notes, with indication of interactive role;
  (v)   List of keywords/ emboldened terms for Glossary;
  (vi) Objectives - keyed to ITQs, SAQs and problems.

Second drafts should contain all the material necessary for the Course Team to evaluate fully
the student workload on that particular Unit, including:
  (i)   Second draft of main text, including Home Experiment notes, AV notes and
         TV/video notes in appropriate style;
  (ii)   Prototype equipment to enable Course Team to test Home Experiments;
  (iii) First draft of AV <u>script</u>, full set of roughs for AV frames;
  (iv)   More detailed plans (or rough script) for TV/ video, with interactive sections fully
         developed;
  (v)   First draft of Glossary entries;
  (vi)   Draft entry for Summary Card;
  (vii) First draft of items for Revision Guide;
  (viii) Additional draft problems (3 or 4) for compilation into the Problem book, with
         solutions.

Handover drafts should contain all the material associated with the Unit, including all
components and contributions to the first year's assessment:
  (i)   Final version of Main Text, including TV/video notes and associated interactive
         exercises, references to AV, and complete Home Experiment notes;
  (ii)   Full set of artwork roughs, including AV frames, ready for artist's briefing;
  (iii) Final AV script (or tape if already recorded) for use by the General Editor in
         checking the frames;
  (iv)   As complete a version as possible of TV/video script;
  (v)   Full set of Glossary entries;
  (vi)   Final version of entry for Summary Card;
  (vii) Final version of entry for Course Summary and Revision Guide;
  (viii) 3 problems for the Problem book, with solutions;
  (ix)   Assignment material:
             * 5 CMA questions, with answers for students;
             * 1 TMA question, with tutor notes and mark scheme;
  (x) . Exam material:
             * 2 computer-marked questions, with answers;
             * 2 short questions, with solutions;
             * 1 long (problem format) question, with solution.

Figure 12.5. Expected contents of the various versions of extended drafts for the
Discovering Physics course.

loneliness is sometimes inescapable, direct experience of it is perhaps also necessary if one is to produce good distance learning material. As he points out (p. 8), 'this essential loneliness also exists for students when they sit down with their course material and attempt to learn'. However, course teams can provide support and encouragement. Importantly too, it is often the sense of responsibility to others and to the shared enterprise, rather than the artificially imposed deadline, that drives authors to produce good work within a limited timeframe.

Riley (1984b) found that the actual procedures by which individuals would prepare 'their' part of the course were rarely if ever discussed, although many authors would privately admit to severe difficulties in accomplishing the drafting stages. Earlier research by the same author (Riley, 1979) also showed that these problems tended to remain constant at an individual level, even through the institution as a whole, and many of its staff, had gained substantial experience of course production over years or even decades.

Apart from time pressures, the most common difficulties that authors face are:

- how to keep students motivated in the right way, and how to ensure that they know what they are expected to get out of the material;
- how to structure the material, i.e. how to thread a logical 'storyline' or sequence of ideas through it;
- how to tailor the package to the amount of study time notionally allocated to it.

Lawless (1994, p. 61) has described distance education texts as 'typically, an uneasy compromise between *telling* and *teaching*' – that is between the provision of basic subject-matter information, and the opportunity to develop and apply knowledge and skills. He ascribes the tension between these two functions to a lack of clear purpose on the part of instructors, which inevitably leads to confusion over objectives on the part of the students:

> Often the overall impression is given that it is the conveying of informa-tion and its absorption by the student which is the predominant purpose ... Even when teaching texts contain exercises and activities they tend to focus on comprehension of the information provided rather than the explicit development of skills.
>
> (Lawless, 1994, p. 61)

Different course teams will find different ways of balancing the tension between 'teaching' and 'telling'. One device used in the *Discovering Physics* course was the setting up of a protocol to help students develop problem-solving skills (as illustrated in Figure 5.8). As well as giving some worked examples, laid out according to this 'preparation/ working/ checking' format, the course team wanted to give students a more informal 'feel' for how a physicist might actually think in trying to solve a particular problem. As shown in Figure

---

WORKED EXAMPLE I

A brick rests on a sloping plank. The angle α between the plank and the horizontal is gradually increased until, at an angle of 40°, the brick starts to slide. What is the coefficient of static friction between the brick and the plank?

PREPARATION

*My figure shows the brick resting on the plank, which is tilted at an angle α to the horizontal. Three forces act on the brick. The weight of the brick, **w**, acts vertically downwards. The reaction force **R** acts at right-angles to the surface of the plank and prevents the brick from sinking into the plank. Finally, a frictional force **f** acts in the plane of the plank, in a direction that opposes the tendency to slide: since the brick would tend to slide down the slope, friction is directed up the slope. The coordinate system has been chosen so that the x-axis points down the slope and the y-axis is perpendicular to the slope. This choice is a good one because the reaction force then has no x-component and the frictional force no y-component.*

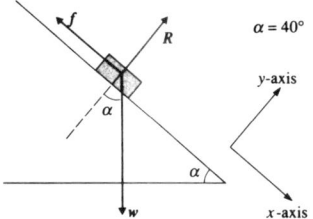

Let m be the mass of the brick.

Let f and R be the magnitudes of the friction and reaction forces.

Gravity and friction can be quantified using laws described in Unit 1. To tackle mechanics problems, you will need to have these force laws at your fingertips.

Relevant equations:

$$w = mg$$

and

$$f \leqslant f_{max} = \mu_{static}\, R.$$

*Ask yourself what is special about the brick. Answer: it is stationary, so its acceleration is zero and the resultant force **F** acting on it must also be zero.*

By the law of addition of forces,

$$F = w + R + f = 0.$$

Figure 12.6. Preparation stage of a worked example in the *Discovering Physics* course, showing the juxtaposition of a formal solution and comments from the author on his approach to the problem.

12.6, this rather conversational, tutorial type of approach was set out on one side of the page, lined up with a formal succinct solution on the other.

Although the sequencing of material will already have been extensively discussed at the outline stage, it is quite common for authors suddenly to become aware of difficulties with the planned sequence as they are developing the material more fully. It is also at the first extended draft stage that it sometimes becomes clear that initial plans were simply too ambitious, and that it is not possible to include everything specified in the outline without overloading the students. Under such circumstances, the author either has to resolve the situation individually, or return to the course team for a further discussion and revision of the brief.

Critical commenting

While many academics are happy to show their scholarship to an outside audience by publishing a textbook, few experience the situation in which their classes are regularly reviewed (still less actually criticized in writing!) by colleagues. Yet critical review is effectively the process that occurs at all stages in course teams. And even the shelter of collective course team responsibility can seem quite flimsy when the finished course is going to be exhibited in its most intimate details to part-time tutors (many of whom will be fellow academics in other institutions), large numbers of adult students from all walks of life, and any member of the general public who can receive the broadcast component.

Course team comments will cover a variety of areas – factual errors, faulty or incomplete arguments, inconsistencies in sequencing, perceived difficulties with the pedagogic strategy, signposting, examples, content that might be added or deleted, stylistic infelicities and typographic slips. All members of the team can usefully contribute to the critical reading phase. Those already expert in the subject matter can check for academic accuracy, and may be able to provide alternative insights into ways of structuring the material. Those less well versed in the details of the content make excellent guinea-pigs: if they cannot follow the argument, find the text too dense, or are baulked by unexplained jargon terms, it is highly likely that students will have the same problems. Such readers are often more articulate than students in pinpointing the source of their difficulty, so the author has a good chance of being able to remedy the deficiency at the next draft. If it is felt that it will not be feasible for the majority of team members to read the drafts, a system of 'designated readers' may be instituted, whereby each person is allocated particular components to comment on, thus ensuring that every draft does receive at least some detailed review.

Comments may be offered at several different levels – emotive response (approval or disapproval), a reasoned argument about what provoked the reaction (especially if it is adverse) and a set of specific suggestions as to what might be done to remedy the perceived deficiency. Even quite serious criticism can be softened by non-confrontational language which puts the author, rather than the commenter, in the position of power ('how about trying . . .', 'some students may get the impression from this that . . .'). Specific suggestions as to how matters might be improved can also help the author to feel that the workload involved in meeting the criticism would not be excessive, and reduce defensiveness. All too often expert readers note down only their criticisms. It is very important, both psychologically for the individual author and for the impact it may have on the overall course design, for commenters also to be specific about aspects that they like.

Another major problem for the author attempting to revise a draft is the receipt of mutually contradictory comments. It is therefore essential that all major issues are aired at a meeting, so that such differences can be resolved in committee. Less important matters may perhaps be left to the author's discretion, allowing him or her to retain some feeling of control and ownership. However, even when something has been agreed at team level, it can be difficult if the authors are complying reluctantly with this decision, as it will then be hard for them to find new formulations that satisfy both themselves and the team. There is a fine line to be drawn here between the imposition of a common style across the course and the professional judgement of each author; this is not dissimilar to the tension the instructors themselves have to resolve between the didactic approach, with its insistence on measurable objectives, and the desire to foster flexible and independent learning. In both cases, one has 'to beware . . . of discouraging creativity for the sake of tidiness. Geese that are capable of laying golden eggs never take kindly to being treated like battery hens' (Butts, 1981, p. 31).

Individual discussions with the most helpful (or perceptive, or academically respected) colleagues may be a very useful addition to the full team discussion. It is also an advantage if the author can receive written comments in advance of the meeting, so he or she does not feel put on the spot in public and has time to prepare a reasoned response to any criticism. It is generally beneficial to allow as long a gap in time as possible between the author completing the draft and the discussion. Apart from giving other team members the opportunity to consider the material carefully, such an interval can also go some way towards distancing the author from the effort he or she has expended, and may therefore make it easier for him or her to receive constructive criticism dispassionately.

Pressure of deadlines, and fear that detailed work may be wasted if course team discussions result in demands for significant changes, may result in some authors of first drafts sticking to mainstream material only, feeling that self-assessment questions, worked examples and access devices can be added at a later stage. However, the more of the intended student-active devices that are omitted from a draft, the more difficult it is for the critical readers to gauge the workload from the student point of view, which should be one of their prime considerations in commenting on draft material. But even if all the components are in place, it can be very difficult for experts already familiar with the subject matter to estimate the time an average student might need to spend on working through a particular piece of distance learning material, although experience gained through evaluation of previous courses can help academics to set their targets more realistically. Length alone is not a reliable guide – students can spend far more time trying to understand something on the basis of a curtailed explanation than they would need to absorb a full explanation written more discursively. Chambers (1989, 1992, 1994) has attempted to develop more objective means of assessing the workload implications of draft course material, and preliminary studies by Garg, Vijayshre and Panda (1992) at the Indira Ghandi National Open University (IGNOU) indicate that they may be successfully applied in the context of distance education courses in science.

Wherever possible, it is obviously desirable for drafts to be produced and read in the same order in which the material will finally appear in the course. This is especially important for subject matter in which sequencing is vital, to ensure that no rungs of the ladder have inadvertently been missed out due to erroneous assumptions about what other authors will cover. However, even when individual modules are less specifically dependent on those preceding them, material generated out of sequence can exhibit continuity problems. The first units to be drafted may receive proportionately more comment. The team members are still fresh, and there is a feeling that what is said about the early parts of the material may have a significant impact on the course as a whole. Later units may receive more cursory reading, as team members become increasingly tied up with their own texts; as already mentioned, the 'designated reader' system can be useful in reducing this problem.

It is generally agreed, as reported by Riley, that two teams of similar size and experience, set the same task of course design with the same resources and target student population would in the end produce substantially different courses. The more innovative the course, the more its final form is dependent on the individuals who have contributed to it. This simply emphasizes the fact that the construction of a multiple media distance teaching course is a difficult, creative piece of design, in which many choices have to be made. These choices are rarely straightforward, being made for a mix of pragmatic and analytic reasons, and as in all design areas, the final product will never be 'perfect'. The evelution of courses as a whole, and of their individual components, is therefore a vital aspect of the continuing process of course creation and improvement. This is the subject of Chapter 14.

4. COURSE TEAM DYNAMICS

Most practitioners who have been critical of the large course team model have nevertheless recognized the necessity of having the process of production underpinned by many people with specialist roles – camera crews, sound recordists, graphic designers, editors, printers, administrators and so on. Discontent with course team operation has tended to focus primarily on the interplay of the academics within the team, and to a lesser extent on their interaction with other team members who may be perceived as having a rather different agenda – especially TV producers and educational technologists. Conflicts between academic priorities and journalistic imperatives, for example, are not uncommon when educational TV programmes are being produced for public service broadcast. Such situations are best defused early on, by agreeing detailed specifications for each course component, and by having a clear understanding about the degree of autonomy allowed to any individual or subgroup in achieving the overall aims of the team.

The extent to which democracy prevails in the course team is at the heart of many criticisms. In the ideal team, all members of the team work by consensus, and a genuine corporate spirit emerges. Some teams can and do operate like this, and they are a joy to be part of. But there are other, less happy, teams with members who perceive the notions of collective responsibility and democratic procedures as a threat. Such people regard 'their' contribution to the course as personal intellectual property, to be defended in the course team against any kind of alteration or take-over which might weaken their rights of ownership, and they can turn every discussion into a confrontation. If they are also strong personalities, their view may actually prevail, sometimes to the detriment of the coherence of the package as a whole.

> The course team format gives the articulate, the domineering and the thick-skinned an influence out of all proportion to their numbers or their merit and conversely weakens the role of the shy, the less verbal and the more sensitive.
>
> (Drake, 1979, p. 52)

However, over-assertive attitudes during the reading and commenting stages can also have the opposite effect. Comments delivered in an aggressive manner may be ignored, even if they actually have academic or instructional validity, perhaps because harshly worded criticisms may be taken personally, rather than being seen as a desire to contribute positively to the end-product.

Virtually all course teams have at some time to face the problems of material that is felt by the majority to be unsatisfactory in some respect, is delivered late, or even fails to materialize at all. Drake (1979), for example, has pointed to the unevenness in style and quality that may be apparent between individual items within the same course, sometimes due simply to idiosyncrasies of style, but sometimes also reflecting variations in ability, experience and commitment between the authors of these different items. Authors who fail to keep to schedule can cause major difficulties, not only for the operations area, but also for their course team colleagues writing later material, particularly in hierarchical subjects like science. Under these kinds of circumstances, the course team chair can be placed in an unenviable position, accountable to the university for the academic and pedagogic standards of the course and for its delivery on time, yet having no real sanctions over team members who fall short of what is expected of them. It is not always easy to be efficient, and yet to involve the whole team in discussion of every single issue that may arise, especially if the team has a large number of members. Delegation is one possible way out of this dilemma. The course might be broken up into more manageable blocks, each the responsibility of a subgroup under a block chair; good integration between the blocks can however be a casualty of this approach. Another alternative is the appointment of an 'academic editor', responsible for ensuring consistency across all the elements of the course. This idea has been used very successfully on science course teams at the UKOU, especially in physics where inconsistencies of notation can cause particular difficulties and might not be spotted by a professional editor without a strong physics background.

Mason (1976) has given an arresting image of the course team chair as a sheepdog attempting to control a flock of skittish sheep:

> The shepherd has a fixed time within which to get his dog to pen the

untrained sheep. It is easy enough to get the sheep started – a few rushes and they are off and running. The problem is to get them running at a steady pace, without baulking when they near the pen. One wrong move and they will scatter. The experienced dog keeps low to the ground, showing himself only when necessary.

(Mason, 1976, p. 28)

In this scenario, the shepherd represents the wider institution and the operations area, whose aim is to pen the sheep as neatly as possible, but who has little actual control over the events.

Although course teams, both individually and collectively, have been heavily criticized in some quarters, they also attract strong support. It is true that there are academics whose primary interest is in the academic content of courses, not in teaching strategy. It is also the case that there are those who enjoy the discussion stages of course design, but are much less comfortable with meeting production deadlines. But it can be quite forcibly argued (see e.g. Costello, 1979) that such shortcomings are more a feature of particular course teams, their individual members, or the way they are chaired or managed, than a structural defect of the course team model *per se*. Others have commented on the sense of common purpose and the creativity that can be achieved through the team approach; for example, Blowers (1979) wrote of the early course team experience of many UKOU academics as follows:

We owe much of our intellectual development and enthusiasm for experiment to the stimulus provided by collaboration, self-criticism, and exchange of ideas fostered within course teams. Some of the courses we have produced would have been unthinkable in the more hierarchical, departmentalized set-up typical of conventional teaching institutions.

(Blowers, 1979, p. 55)

13

THE ROLE OF THE TUTOR

Of a good leader, when his task is finished, his goal achieved, they say,
'We did this ourselves'.

(Lao-tse, *c.* 600 BC)

1. SUPPORT FOR THE STUDENTS OR TEACHING FOR THE COURSE?

Most learning is, as discussed in Chapter 3, a socially constructed activity. Although there are a few students who choose a distance learning course precisely for its element of distance and prefer to work entirely remotely from others, the vast majority want or need direct contact with experts and with other learners following the same programme as themselves. For some, occasional 'troubleshooting' help, of the sort that could be supplied by a brief telephone conversation or an exchange of letters with a subject specialist, may be sufficient. Many students, however, require more regular and structured contact with someone who has the expertise to bridge gaps between them and the course material, and who can respond flexibly to their individual difficulties and questions. Most learners also value peer group interactions, perhaps mediated by an expert facilitator, as a means of sharing ideas and of checking their own understanding and progress. Group meetings can generate a feeling of belonging to a wider student community, and this of itself can boost both confidence and motivation, especially in learners who otherwise study in isolation. The phrase in commmon use in distance education to describe the interactions involved is 'learner support', which is normally understood to cover both content-specific and learning-specific issues, and is focused on responses provided to an individual, known student. In that sense, learner support is quite different from

> interactive devices embedded in materials which, however successful they may be in generating a personal response from the learner, cannot respond to that learner as known individual, during the process of learning itself – the learner can only be known as an abstract entity before learning begins.
>
> (Thorpe, 1994, p. 148)

Different distance learning institutions have devised different systems for learner support. In some, the specialist and generalist roles of facilitators are separated, even to the extent of being vested in different individuals with different titles, such as 'tutor' and 'counsellor'. In other systems, the roles are less sharply distinguished, or even merged. Thus in the UKOU, 'tutor-counsellors' support Foundation level students in all aspects of their learning, providing specialist academic interpretation of the course material in response to student needs, and establishing an individual relationship with students that facilitates their learning; the tutor-counsellor thus mediates a student's interaction with the course material, the other students in the group and the university community in general. In most distance education systems, an important component of the tutor's function is to comment on work that the students submit to them, and usually to grade it. Section 2 considers this 'correspondence tuition' role in more detail. The tutor may also be expected to run face-to-face sessions with students, on either a group or an individual basis, and these are the subject of Section 3. Throughout the chapter, we will use the term 'tutor' in its broadest sense, to describe someone whose role is to provide academic tuition to each member of a small(ish) group of learners, to give them individualized feedback on, and to grade, the work they submit as part of the continuous assessment system, to run group meetings, and to provide general support in furthering their academic progress. This last function is often the reality of the matter anyway, though not necessarily the way the system is set up: even if the tutor does not have a formal 'counselling' role, he or she is the person to whom the students most naturally turn if they have problems, academic or otherwise, that affect their studies. A comment from a UKOU tutor underlines this point:

> It's difficult in the OU to distinguish when you're counselling from when you're tutoring; usually you're doing both at once . . . If you're on the spot, and you know the answers, then you're the one that gets asked. And you're the one who gets told about holidays, house moving, new jobs, illnesses, and all the other crises which make assignments late.
>
> (Open University, 1988, p. 43)

The tutor may well be required to clarify parts of the course material that students have found difficult, encourage them to engage actively with the material and help them apply what they have learned in different situations of the tutor's devising. It is therefore essential that he or she have good academic qualifications in the subject-matter of the course. However, tutors must appreciate that their role is not primarily the delivery of content, but the fostering of learning. The content is provided by the course materials. What students mostly require from their tutors is help in unpacking the skills necessary to handle that content. Good tutoring practice in this respect can come in many forms, but important criteria for success include:

- a reasonable regard for the content and teaching stance of the course material;

- an awareness of the main blockages to learning that can occur in the subject area, and the ability to devise ways of overcoming them;
- an enthusiasm for the subject of the course, and the ability to convey this enthusiasm to students;
- the ability to create a non-threatening, stimulating atmosphere that is supportive of learning;
- a willingness to offer opportunity and assistance to students that will help them develop general and subject-specific study skills appropriate to the level of the course;
- an interest in encouraging academic progress, on an individual basis and for students with a wide range of backgrounds, abilities and motivations;
- an openness to new ideas and teaching methods, and an active approach to participation in training and professional development opportunities;
- an appreciation that one of the measures of success is the extent to which the students become 'independent learners' (that is independent of the tutor!). A UKOU tutor has described his thinking on this aspect of his work as follows:

When I talk about helping students to find solutions to problems I don't mean by that a situation in which *I* show *them* the answer. I *do* mean by that a situation in which I help them to find that answer for themselves, and to be better able to find similar answers in similar situations in the future, without my assistance.

(Cowan, 1992, p. 6, original emphases)

An important consideration from the institution's point of view is the fact that, though many tutors are selected not only on the basis of their subject expertise but also *because* they have teaching or lecturing experience from other sectors, this experience may not be directly transferrable to the distance education setting:

Organising and supporting pupils' learning in schools is the essence of the teacher's role. In a university department the role changes [with a] . . . narrowing of teaching responsibility to areas of specialism, rather than the broader remit of general learning support.

(Partington, 1995, p. 12)

In many systems, tutor posts are part-time, and may be taken up (often very successfully) by people who have no formal training as educators, such as those from research establishments, women working from home, and the increasing number building 'portfolio careers'. Some will have prior experience of counselling, others will have none. The diversity of backgrounds of tutorial staff, and the special nature of their role, puts a particular onus on the distance teaching institutions to provide proper training and staff development opportunities. This is an issue we will examine in more detail in Section 5.

One way to summarize the tutor's function is to say that it is to provide the 'human face' of the institution. This is seen by many as a pivotal role in ensuring that as many students as possible stick with their courses, get what they want from their studies, and develop their full potential: 'The importance of personal contact and relationship building remains a critical component of most effective distance education programmes and is likely to increase as technical capabilities expand' (Wills, 1994, p. vi).

2. CORRESPONDENCE TUITION

Although in any distance education system that is heavily reliant on printed materials sent through the mail there is a sense in which students receive the bulk of the instruction 'by correspondence', we use the term 'correspondence tuition' here to mean the written communication between tutors and students in the form of feedback on set work. In many institutions, this is the only form of individualized tuition that is available to every student, given that face-to-face sessions, if they are arranged at all, may not be accessible for reasons of geography, timing or disability. In what follows, we assume a pattern (based on that of the UKOU) in which the assignments are actually set by the instructors responsible for the course material, but are commented on and graded by the student's own tutor; this tutor may also meet some, but not all, of the students on his or her 'list' at tutorial sessions.

If the assignment is well designed, it will be an integral part of the learning process. The students will not be able to complete the task satisfactorily just by regurgitating material from the course, but will have to restructure their knowledge and apply it to new situations. The tutors then have a dual function. They *comment* on the students' efforts, not only correcting, redirecting or expanding where the answers are lacking in some way, but also praising what is good and hence reinforcing the learning experience. Secondly, they *grade* the work, according to a set of criteria laid down by the originators of the assignment.

Given that correspondence tuition may be the *only* form of individualized teaching that the student receives, it is this that sets the tone for the relationship between student and tutor. Herein lies the art of the good correspondence tutor, who has to develop entirely through the written word the kind of communication that would more normally in educational circles be conveyed by a combination of written notes and spoken exchange. Each student will probably require a different array of comments – for one learner the problem may be with understanding a basic concept, for another it may lie in mastering the presentational skills necessary to demonstrate their understanding, for another still it may be in selection of the material required to make a fully reasoned argument. The tutor's annotations have to encompass (no hierarchy being implied by the ordering of items):

- comments that establish the dialogue, recognize the validity of the student's own efforts and encourage further contact;

- comments that correct errors of fact or flaws in logic (such comments are vital, since leaving errors uncorrected may implicitly reinforce misconceptions) and comments that try to disentangle the root of the student's misconception about something – without making him or her feel stupid or inadequate;
- comments that address significant omissions or irrelevancies in the student's answers – without making him or her feel the time spent on the assignment was wasted;
- comments on structure, style, presentation, layout of calculations, etc.;
- comments that praise what is good in the student's answer, and invite him or her to extend this thinking, building on areas in which success has already been achieved;
- comments that encourage the student to keep going, and renew his or her commitment to the course as a whole and to the next assignment in particular;
- comments that explain the reason for the grade, set the mark in the context of the student's overall progress, and indicate how this grade may be repeated or improved upon in future assignments. Ensuring that the mechanism of grading is clear to the student, and therefore also serves a teaching function, can go quite a long way towards resolving the potential conflict that may exist for the tutor who is simultaneously the facilitator of a student's learning and the judge of that student's progress.

Figure 13.1 illustrates a number of cursory, 'closed' statements that might be made about students' written work, and some possible alternatives that are more successful in fostering a learning atmosphere.

3. FACE-TO-FACE SESSIONS

It is precisely because face-to-face tuition is a scarce and relatively costly commodity in a distance education system that it is important to question and justify the processes involved in it. Institution, tutors and students all need to understand the underlying reasons for face-to-face sessions, what tutorials are expected to achieve, why particular techniques are deemed superior to others, and whether this form of student support is educationally effective and 'value for money'. Figure 13.2 summarizes some of the functions of tutorial provision in a distance education system.

From the students' point of view, a tutorial must enhance both their learning and their interaction with the course material. They will not see attendance as worthwhile unless it contributes more to their progress than would the same amount of time spent studying on their own at home. A lecturing, or largely tutor-driven, approach will almost certainly fail to satisfy this criterion, because it represents a very similar type of experience to that obtained by reading a text

Closed/ negative/unhelpful comment	Comment designed to help/encourage the learner
No! See model answer attached	Unfortunately, you have missed the point here. I've attached a specimen answer to set you on the right lines - look at this in conjunction with pp. xx of Unit Y, and do contact me if you are still not happy with the topic.
This line doesn't follow from the previous one *(no indication of actual mistake)*	You have a made an algebraic slip here. Remember that the factor 3 multiplies <u>both</u> the terms inside the bracket, so this line should be $3x+6$. I have taken off one mark for this mistake, but have not penalised you again for the resulting propagated errors.
Careful! adaptive resistance is not the same as immunity	You may have confused two concepts here. Bacteria don't become immune to antibiotics as you might become immune to measles. Adaptive resistance is entirely due to mutation (already present in the population) + selection (due to antibiotic), leading to evolution of resistant strains.
Well over the word limit - stick to the constraints imposed by the question	Compare your first paragraph with my version which I've attached on a separate sheet. Yours doesn't actually say any more, but it is four times as long. Note that I've made no attempt to <u>explain</u> concepts in my answer - the question only asked for <u>description.</u>
True, but why?	What you say is correct, but you were expected to back your statements with explanations/ reasoning.... For example....
Padding/ repetitious/ irrelevant	What you say is largely correct, but the sentences I have highlighted are not central to your main point/ repeat points you have already covered/ are not strictly relevant to the discussion of .... (though they would have been relevant had you been asked about....).
These two molecules are not structural isomers - they're <u>the same.</u>	Actually, these two are identical. You may be able to see this if you imagine 'turning over' molecule A. But it can be difficult to visualise the molecules just from the structural formulae drawn on the page. You might find it helpful to construct a molecular model using the kit to convince yourself.

Figure 13.1. Correspondence tuition: phrasing comments constructively.

or listening to a broadcast. Indeed, any tutorial plan that is based principally on the transmission of information is fundamentally inappropriate in that it sets up the tutor as the supplier and the students as the recipients of knowledge; this may boost the tutor's ego, but does very little to help the students in taking responsibility for their own learning. The tutor has to submerge his or her own inclination to put on a polished 'performance'. Instead, the agenda for the session has to be negotiated with the students, and allowed to develop in accordance with their needs.

> Teaching that puts learning first looks at first sight less impressive [than a presentation by the teacher] . . . It's often slower, with many silences. Students talk to one another rather than answering the teacher's questions . . . When questions are asked by the teacher, they are genuine questions, not a version of 'Guess what I'm thinking'. They ask what the student is thinking, what premises are being used. Testing how much the student knows is done some other way (work-sheets, reviews of the units), not by an abuse of the Socratic method . . . The devices are laid bare; this is the least manipulative kind of teaching.
>
> (Open University, 1988, p. 11)

Attitude and personal style are just as important in generating a learning atmosphere as are 'macro' methods of structuring a session:

> He'll be careful not to finish the student's sentence. He'll say 'Yes, go on.' There is a low word count coming from him. He's very patient – he doesn't say 'Oh, I'll tell them and get on with it'.
>
> (Gibbs and Durbridge, 1976, p. 11)

The real *raison d'être* of the tutorial is to allow for dialogue between individual students and the tutor and also, importantly, between students themselves. Unless there is dialogue, the tutorial will offer little that could not be achieved equally well by sending the students an audio or videocassette. Some teachers, and indeed students, occasionally question the need for dialogue and debate in a subject such as science, which may be viewed as based largely on fact and logic, and in which there are are 'right' (or at least generally accepted) and 'wrong' answers. However, learning science is in fundamental respects no different to learning in any other area; one still has to be able to articulate one's difficulties and ideas, and discuss them with others engaged in the field. A true understanding of the processes of science is not gained just by reading, watching or listening – it is acquired by questioning, hypothesizing, testing and

---

What is the purpose of tutorial sessions?
- clarifying and consolidating course content
- helping students overcome conceptual misunderstandings or blockages
- applying skills and concepts covered in the course to different examples
- developing study/learning skills
- providing group support/learning, sharing experiences
- building students' confidence (in tutor, the group, themselves)
- establishing a series of regular contacts (an aid to perseverence and pacing)
- providing opportunities for remedial tuition/ individual counselling
- initiating self-help pairings or groups

Figure 13.2. Face-to-face tutorials: what is the purpose of the sessions?

doing. The tutorial situation allows students to participate in these activities in ways that cannot be exactly replicated by distance teaching material.

This is not to say that a tutorial is necessarily educationally valuable simply because the students take an active role in it. The tutor has to do more than simply create situations in which some sort of activity will occur. The students must gain some demonstrable educational benefit, and this will only happen if learning objectives are considered by the tutor at the planning stage and the activity structured in such a way as to foster the achievement of those objectives. There also has to be a *variety* of activities, to allow for the whole spectrum of student difficulties and motivations, and to ensure than the session moves forward with due regard for concentration spans and periods of consolidation.

Despite this emphasis on group activity, the tutor also has to satisfy the individual needs of students who are having difficulty with the course:

> learning can be totally blocked – often in utterly frustrating and highly personal ways. It has certainly been my experience that many learners . . . can come to a tutorial with a shopping list of such blockages – which have frustrated any progress whatsoever on their parts. So I would say that, in planning a tutorial, a tutor should aim to provide assistance in unlocking the 'log-jams' which are frustrating the learners
>
> (Cowan, 1992, p. 6)

Figure 13.3 is a checklist of issues that might usefully be considered by a tutor in preparing for the face-to-face component of his or her teaching. (This list is not intended to be exhaustive, but rather to act as a starting point.)

There is a natural desire on the part of both tutors and institution to find out 'what works best' in terms of tutorial provision generally and in terms of the structure of individual sessions. However, in a system in which tutorial attendance is voluntary and may also be genuinely inaccessible to a proportion of the students, it is extremely difficult to separate all the variables involved, in order to derive valid quantitative information. For example, the mere fact that Tutor A consistently draws more students to his sessions than Tutor B, who teaches the same course at the same venue, might be due to any combination of the following (or other additional) circumstances:

- Tutor A is more adept in serving the needs of his students, or runs more interesting sessions;
- Tutors A and B are equally adept, but Tutor A has a stronger group of students, who welcome tutorials as an opportunity for further discussion, in order to widen/deepen their ideas;
- Tutors A and B are equally adept, but Tutor A has a weaker group of students, who perceive themselves to be more in need of face-to-face tuition;
- Tutor B has a larger proportion of students in his group who cannot attend sessions because of difficulties of transport, care of dependents or work patterns.

There will be further variables to consider if one attempts to compare tutorial

---

What are the main issues to consider in planning a particular session?
- taking account of time of year/ stage of the course
- establishing tutor's own objectives for the session
- knowing the setting (layout of room, availability of visual aids, special arrangements for students with disabilities, etc.)
- planning for variety
- planning activities that will cater for the spread of ability within the group
- having a flexible plan that will allow for negotiation of the agenda with the students on the day, and can be adapted according to the number of students actually present, their responsiveness and needs
- balancing content-specific and skills-oriented aspects of the session

What are the key factors in actually running the session?
- icebreaking/ introductions
- agreeing the objectives/structure of the session with the students at the beginning
- helping students to admit/ articulate their difficulties
- using a variety of media, e.g.
    * 'chalk and talk', OHP, worksheets, rock samples, molecular models,
             past assignment or exam questions
- using a variety of techniques, e.g.
    * lecture, demonstration
    * brainstorming, plenary discussion
    * pairing, snowballing, pyramiding, buzz groups
    * games, simulation, role play
    * concept mapping, building flowcharts or grids
- taking account of the group dynamics (e.g. coping with dominating/ reticent students)
- ensuring participation by all
- balancing the needs of individuals with those of the group
- summarising the main points/ achievements at the end

How should the session be evaluated?
- reflective practice, e.g.
    * what did the tutor feel was the most effective part of the session? why did it work?
    * were any of the objectives for the session not achieved? why not?
    * did all students participate?
    * which activities did the students enjoy most/appear to learn most from?
    * were there any questions/difficulties that the tutor had not anticipated?
- obtaining feedback:
    * on what aspects would it be most useful to have feedback, and how might it be obtained (e.g. from students, peers, line managers)?

Figure 13.3. Planning a tutorial: a checklist of issues.

attendance across different courses, locations or catchment areas. Ultimately, each tutor has to answer for him or herself the question 'did that particular session work well for those particular students at that particular stage in that particular course?' The methodologies ('reflective practice', action research or student evaluation) by which this kind of question may be tackled will be explored further in Sections 4 and 5.

In advocating a learner-centred approach, one should not underestimate the extent to which novice tutors and students alike may be uncomfortable with the idea of 'tutor as facilitator'. From the point of view of the tutor, the feeling that tutorial time is so limited and there are so many essential points that students *must* grasp to do well on the course can lead to a tendency to talk too much in a well-meant attempt to 'give students their money's worth'.

(Open University, 1988, p. 55, original emphasis)

Welding new students into a group supportive of one other can also sometimes be an uphill struggle. Some members of the group will be anxious to bombard the tutor with their own ideas and questions, or to dominate the group with their demands. Others will be reluctant to contribute anything themselves, for fear of looking foolish. Others still may expect to obtain from the tutor what they perceive as expert advice and 'instruction', and hence be unwilling to listen to, or to work with, their peers. Time is needed to reconcile or satisfy these varied needs and expectations, and time is usually very much at a premium. If the students meet infrequently, as is common in distance education systems, they may find it difficult to develop the relationships and trust between members of the group that are conducive to collaborative learning. The tutor may have to work hard at convincing them of the pleasures and benefits of the collective learning power that they can themselves generate. And it is ironic that one measure of the tutor's success may be the enthusiasm with which the students eventually form self-help groups, the first step on their road towards becoming independent learners.

4. TENSIONS IN THE TUTOR ROLE

The 'tutor as facilitator of learning' approach can create a number of conflicts for the staff involved, which need to be resolved if the role is to be a satisfying one for the individual, and ultimately a successful one for the institution. In a distance education system, the curriculum and pacing are pre-defined by the institution, as are the assignments that the students are set and the mark schemes according to which these assignments are graded. There is usually even detailed guidance on how to run the tutorials. What is left for the tutor in terms of creative input? Another kind of tension arises from the fact that most tutors regard the face-to-face sessions as by far the most enjoyable aspect of their work, and yet this is but a small part of their overall job description. A third source of conflict, already noted in Section 2, can be the dual responsibility of the tutor, as both teacher and academic counsellor of the students, and as the assessor of their performance. Added to all this, there is the problem of remoteness from institution, course designers and peers that tutors share with their students – what Kirk (1976) has referred to as 'the loneliness of the long distance tutor'.

The first question, 'what is in it for the tutor?', is to an extent answered by pointing out that if learner support has any effect at all on students' progress – and there is plenty of evidence to suggest that it does – then this can be attributed directly to the activities of the tutorial staff. The personal satisfaction in teaching self-motivated, adult students can be tremendous, especially as they are often more obviously appreciative of their tutors' efforts than are younger learners.

> There is a genuine intellectual tension in teaching a course which you have had no hand in making, but it is a creative one. You are freed to do your best teaching and you and your students are more nearly equal partners in the enterprise.
>
> (Open University, 1988, p. 20)

The student motivation factor is a crucial one, however. It is probably the principal reason that direct contact with the students who come voluntarily to tutorial sessions is usually seen by tutors as the most interesting and enjoyable side of their role. 'Proactive' contact – such as chasing up a student who has failed to submit an assignment on time, who seems to have made an inappropriate choice of course, or who needs special arrangements to cover a short posting abroad – is a much less attractive facet of the job. Yet there are strong suggestions that it is the proactive, academic counselling activities that have the most effect on individual students' progress and therefore ultimately on the institution's success in terms of student retention and throughput.

In a system that tends to emphasize above all the learner support aspects of the tutor role, tutors also have to feel that the institution recognizes and values their individual academic expertise. One way in which tutors can derive satisfaction in this respect arises from the difficulty, noted in Chapter 12, of keeping course material up to date when it has to be prepared well ahead of the first presentation of the course, and is expected to stand throughout the designated course life (maybe as long as a decade). Tutors can play a valuable role in bringing to students' attention recent advances and shifts in emphasis, and in helping them to appreciate that the process of science is one of continuing debate. This obviously needs to be done sensitively, so as not totally to undermine the material on which the learners are basing their studies and on which they will be tested, but provided students do not feel that their progress is being threatened by the introduction of 'extraneous' material, they are usually receptive to a tutor who can present a brief account of up-to-the-minute developments. And this can be a positive avenue for tutors who may otherwise feel that their contribution is too constrained by the strait-jacket of the course content and structure. One UKOU tutor has written that his attempt to put across the current debates about the biological bases of behaviour

> gave me an opportunity to be more than a mechanical scrip-marker-cum-remedial-problem-solver: it gave me the feeling that, in my own small way, I *was* able to contribute directly to the education of my students and the development of their understanding of science, to impart a little of my own accumulated wisdom.
>
> (Dunbar, 1982, p. 70, original emphases)

Over a period of time an enormous reservoir of expertise builds up among the tutorial staff in how to make their subject exciting, interesting, relevant and accessible to typical students. Tutors also get to know, through direct contacts not available to the course designers and authors, what parts of the courses students find difficult, overloaded or dull. It can do a great deal for tutor morale if the institution can find ways of drawing on this expertise, so breaking down the gulf that is often perceived to exist between the 'ivory tower' academics who produce the courses and the 'workers at the coal face' who mediate students' interaction with the material. As mentioned in Chapter 12, it may be possible to involve some tutors in course preparation, perhaps as critical readers or as consultants. Tutors can also play a vital part in on-going review and evaluation of courses in presentation. They may be able to contribute to the production of assignment questions, which is a task that is usually undertaken afresh for every presentation of a course. Tutors are the right people to design tutor resource packs, which can help to disseminate innovative ideas and good practice among their colleagues. They may also be able to reduce their own remoteness by participating in peer group activities – such as mentoring new tutors, monitoring colleagues, etc., and to derive considerable personal satisfaction simply from honing their own performance in the role. The value of such staff development opportunities will be further examined in Section 5.

5. STAFF DEVELOPMENT AND QUALITY ASSURANCE

It is rarely possible for distance teaching institutions to recruit tutors with prior experience of distance, or even open, education. Newly appointed tutors therefore normally have to go through a period of briefing and training, during which time they find themselves on a very steep learning curve. They have to penetrate the jargon of the institution, discover how to navigate its administrative structures (both on their own behalf and on that of their students), check in detail the academic content of the course they are teaching, get to grips with the usually quite unfamiliar demands of correspondence tuition, and prepare for meetings or telephone contact with students. The priority during this induction period is therefore to give tutors the essential information about the way the institution operates, and to provide them with a core set of skills and techniques that will enable them to support their students most effectively. If, as would normally be the case, the tutor is well qualified academically to teach up to or beyond the level of the course, the initial training can be virtually non-subject specific, concentrating instead on the *processes* of tutoring.

However, it is important that the sharing of experience, teaching strategies and good practice should not be confined to an induction period, but should be an on-going part of tutors' professional development. As Lentell has noted, this is for the institution 'a significant component of the quality infrastructure – the means to get beyond merely reproducing its practice but to research and develop it' (Lentell, 1994, p. 51).

Concerns have been expressed (see e.g. Partington) that many such programmes in the higher education sector have been driven more by externally imposed quality assurance audits than by any intrinsic motivation of staff to develop their teaching in response either to reflection on their own practice or to the changing needs of their students. In fairness, this is certainly not true of all programmes or institutions, and is perhaps least true of the distance teaching institutions, where the 'unusual' nature of the tutor's role has always been recognized and supported by good briefing materials. However, with changing course contents, and a regular influx of new tutors to be trained, there should be even in the established distance education universities a constant review of staff development opportunities and processes:

> One of the major concerns of all staff is with preparing students for their changing and challenging futures. The ability to learn, to continue to learn and to be amenable to change are key to that preparation. One of the most successful ways of developing those habits and attitudes in students is to ensure that they are reflected in our own behaviours and practices.
>
> (Partington, 1995, p. 12)

Figure 13.4 summarizes some of the interactions in the circle of critical reflection that tutors can bring to bear on their own teaching. The processes of review, learning from students and adaptation should be an integral and continuous part of all teaching. Unfortunately however, evaluation for personal development can all too easily be misused as evaluation for judgement of performance, leading in turn to defensiveness on the part of the tutor. Part-time tutors in a distance education system tend in any case to feel more exposed than teachers in many other sectors. Their comments on students' assignments have to be written out for all to see, they are more likely to have full-time staff sitting in on their classes, and their performance is constantly being assessed by highly motivated adult learners with a wide range of experience. The need for standardization of the continuous assessment procedures means that their correspondence tuition and grading is compared directly with that of other tutors on the same course. Peer group discussion offers a way of removing much of the threat of all this scrutiny, and turning it to positive advantage. Meetings billed as 'staff development' should give tutors opportunities to discuss in a critical but non-judgemental way their practices and attitudes to the role.

> We learn from one another's good practice, from the similarities of our failures, and from the analyses of our successes which we have to make in order to explain them to colleagues – much as the students learn from one another's attempts to learn.
>
> (Open University, 1988, p. 19)

In a distance teaching organization, peer group dialogue, like student–tutor dialogue, need not depend on meetings, but can extend into correspondence. For instance monitoring (i.e. quality control checks) of correspondence tuition may be carried out by those responsible for running the course at the institutional level, thus providing valuable feedback to the questions setters which can inform their construction of future student assignments and a direct link between them and the tutors. Valuable too – though in a different way – is monitoring by other tutors, which has been found to have benefits even beyond that which might originally have been envisaged. Those monitored often feel more comfortable with the process when it is carried out by their peers, who may be supposed to share their perceptions of the system and do not have any line management responsibility towards them. But equally importantly, those doing the monitoring almost invariably report that it is a positive learning experience for them: the opportunity to see the range of approaches adopted by others results in an improvement in their own correspondence tuition.

For any staff training and development programme to be really effective, its relevance must be demonstrated to and embraced by the staff involved. This may seem a truism, but tensions can certainly exist, for instance for tutors

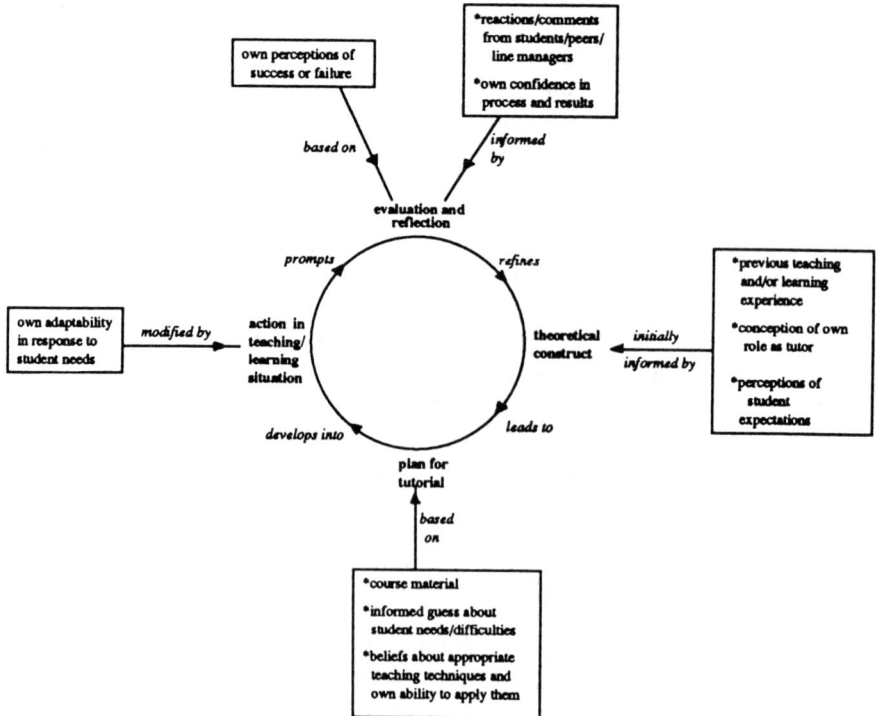

Figure 13.4. Circle of reflective practice in tuition.

whose teaching styles do not match institutional expectations. Similar difficulties can arise insidiously with long-serving tutors, whose approach to teaching may not keep pace with institutional evolution, or who may simple become 'stale'. Even those who are apparently keen to participate in peer group discussions may seek in essence to restrict the agenda. The interaction between teacher and 'class' is often seen as a private one, and this means that while most teachers are happy to debate issues of content – level, sequencing, etc. – they can quickly become extremely defensive about discussions of process, especially when these move on from mechanisms (e.g. worksheets) to more personal matters such as group dynamics. There is no sense in *telling* tutors that facilitative learning is the system imposed by the institution. One important aim of staff development meetings should be actually to *use* small-group methods to give tutors a very positive experience of the power of the technique; for this to have maximum impact, the topic of the session should not be group work itself, but some other subject of genuine relevance and importance to the group. Northedge (1987) has highlighted the effectiveness of role play in this regard:

> If one models a classroom situation, using teachers in the role of students, they immediately begin to draw (very creatively often) on [their] large fund of experience from student days. Memories of hostilities and boredom, stupefaction and competetive aggression become manifest in the acting . . . what is more this happens collectively, so that when the role-play is discussed afterwards one has a whole group of people who have communally crossed the barrier that normally separates past student experience from current teaching experience. What is normally routinely ignored or denied is brought powerfully into the arena, but not in a threatening way.
>
> (Northedge, 1987)

By encouraging tutors to empathize with the experience of their students, and by actually demonstrating the benefits of interactive techniques for the tutors themselves, these kinds of staff development sessions can bring the reflective processes outlined in Figure 13.4 more vividly to life than any amount of solo analysis.

14

EVALUATION

No course is perfect, no course will ever be perfect. But, if in the light of student performance and comment, amendments can be made to course material, the students of later years will find an improved course.
(W. Perry, UKOU Vice-Chancellor's Report, 1969)

1. INTRODUCTION

The efficacy of course design eventually needs to be tested. In this chapter we describe the range of ways in which information can be collected and interpreted to discover how course designs have worked out in practice. Evaluation involves such diverse activities as market research (i.e. consulting potential students when courses are in the initial planning stages to gauge possible enrollments), developmental testing of courses in preparation with volunteer students, collection of feedback from students on courses in presentation, and analysis of performance data on students' progress. Two sorts of background contribute to an understanding of work on evaluation, a historical perspective on the evaluation methods in education and applied psychology, and the educational technology approach. The next two sections of this chapter will deal with these two aspects.

2. TYPES OF EVALUATION

There is a long history of inventing and refining methods in educational evaluation. The single most important consideration in designing the evaluation of an educational initiative, whether a series of courses or a single short example of instructional design, is the *purpose* of the evaluation. Evaluations can be carried out for teachers or planners in order to decide whether a course should continue, or whether more similar courses should be planned. Evaluations can be carried out for course designers to decide how to improve a course from the learners' point of view. Evaluation can be carried out for a funding agency for a wide range of reasons. Evaluation can also be carried out by external agencies such as professional associations, with the aim of providing information for employers, or even for other education providers, about the skills, knowledge and personal qualities that learners have developed in a series of courses. Another purpose of conducting an evaluation might be to quantify cost effectiveness, yet another might be to discover whether a course is successful in attracting students of a particular type. But as Calder (1994) notes, it is likely that there will be some common features of interest whatever the main purpose of the evaluation.

> All stakeholders are . . . likely to be concerned about the student survival and pass rates, about whether the course reaches the standards agreed with its accrediting body, and whether the students find their learning experience a satisfactory one.
>
> (Calder, 1994, p. 196)

There are a variety of ways of conducting an evaluation. The issues can best be summarized by presenting types of evaluation as polarized dimensions, e.g. qualitative/quantitative, formative/summative, etc. The qualitative/quantitative contrast is obvious. There is certain attraction in focusing on those aspects of an evaluation that can be measured and expressed in terms of numbers or percentages, but there is a danger that those aspects which are difficult to quantify (such as students' enjoyment or orientation to study) will simply be neglected. The next pairing of formative/summative focuses attention on the purpose of the evaluation. Formative evaluation is conducted so that the results from it can have an impact on a course design before it is completely realized. Summative evaluation occurs at the end of a project to assess its success. Scriven (see e.g. 1980) first made this distinction, and it is discussed in Kandaswamy (1990).

The next dichotomy is illuminative/controlled evaluation. This division first emerged in work by Hudson, famous for his discussion of subjects like creativity (e.g. 1966), and attention to the distinction encourages evaluators to consider non-traditional sources of evaluative information, rather than simply considering the inputs and outputs in a learning situation. Parlett and Hamilton (1972) developed the illuminative approach and concentrated on developing an understanding of the processes which students undergo while studying a course. A closely related approach is that of Stufflebeam (1971), who considered the process that takes place during the presentation of a course, as well as measures of the intended outcomes and a consideration of the context in which the course is being used. Kemmis (who had been Hudson's student) made the distinction between ideographic and nomothetic evaluation approaches. Ideographic approaches involve very open exploration, whereas nomothetic systems are usually strongly rule-based and involve commitment to a particular set of values. Thus in a nomothetic evaluation systems are measured against these values. Curriculum evaluation has been dominated by versions of one or other of these two models. For example, MacDonald, with Kemmis a main protagonist in the UNCAL (UNderstanding Computer Assisted Learning) evaluation based at the University of East Anglia, produced ground breaking evaluations of the National Development program for computer-assisted learning in the 1970s. MacDonald has also written about the importance of conducting longitudinal rather than snapshot evaluations. Early evaluations tended to be quantitative and controlled, but in recent years the trend has been more towards the qualitative, illuminative and ideographic.

Another way of describing some of these differences in evaluation models is to distinguish between an objectives, or testing, school of evaluation typified by that of Tyler, and an alternative school associated with anthropological models (Parlett and Hamilton) and connoisseurship (Eisner). A relatively recent distinction is that between context-based and decontextualized evaluations, which arises in part from the situated cognition research paradigm mentioned in Chapter 3. Workers on developmental psychology, cognitive science and anthropology have joined forces in a number of projects to examine how learning is influenced by social context (see e.g. Rogoff and Lave, 1984).

Choice of the appropriate evaluation methodology is strongly influenced by the model of learning adopted. The possibility has been suggested (by e.g. Papert, 1980) that the most important types of learning are difficult to evaluate for a number of reasons, chief among which is that the learning may not manifest itself immediately. Studies of collaborative computer-based learning in science (e.g. Scanlon *et al.*, 1993; Issroff *et al.*, 1994) have established that a complete picture of the learning outcome sometimes only emerges from considering the students' performance some time after the period of instruction, as a result of delayed testing of the students. This phenomenon is alluded to in the following quotation which examines the model of learning underlying certain approaches to conducting evaluations:

> In a nutrient like model of the teaching and learning process, you expect learning (growth) to occur best or at all only if multiple factors are all present and present over a considerable time. Adding a pinch of nitrate, selenium or whatever seldom causes a plant to leap upwards overnight. So even if a piece of teaching does directly contribute to learning, the effects are likely to be slow; and it will only occur in the presence of many other co-factors. This implies both that measurements should extend over a long period, and that we cannot expect to understand results unless we track many possible factors . . . Only highly artificial experiments that completely control an organism's access to the separate components of nutrition over longer periods can really establish these factors. When you know what they are, then you can analyse any input . . . and so say what a particular input is contributing to the eventual outcome . . . Evaluation then becomes an analytic process on the inputs, not a matter of direct empirical measurement of the effects. A weaker conclusion for evaluation would be that measurements must include not just learning outcomes, but the value of the various factors thought to be important.
>
> (Draper *et al.*, 1994, p. 32)

Another component of the design of an evaluation strategy is that sufficient weight needs to be given to affective aspects of the learning. As we mentioned in Chapter 3, a number of writers are beginning to address the relationship between cognitive and affective factors in the learning of science. Motivation and the student's perception of the purpose of certain learning tasks are both crucial to this relationship. Motivation could be thought of as one of the factors referred to in the preceding quotation. It is necessary to ensure that the motivational structure is the same in the evaluation experiment as the learning situation, so the fairly common procedure of paying students to work through draft instructional material and fill in evaluation instruments is suspect in this regard.

The conduct of an evaluation requires not just an understanding of the space of possible types of evaluation but requires the evaluator to construct an evaluation plan. It is tempting to start designing a plan by itemizing the sorts of data to collect, but this approach can cause problems if no consideration is given to a wider range of questions. The list below is one version of those questions.

- What is the purpose of the evaluation?
- What to evaluate?
- Who should conduct the evaluation?
- How to evaluate?
- What data are worth collecting?
- What methods can be used?
- What reports are necessary?
- How will the audience receive the results?

Stake's answer to the first question describes three possible foci of what to evaluate – antecedents or aims, transactions or processes, and outcomes or results (Stake, 1967, 1978). In each case he believes that data need to be collected on both intents and observed effects. The question of who should conduct an evaluation has been a subject of debate. The basic problems are that an evaluator distant from the course producers can be seen as a threat, while an evaluator too close to the producers is seen as not sufficiently impartial. The methods used to overcome these difficulties are best described in the context of the specific examples we present in the next section, but, as will be seen, we favour an eclectic approach to methodology.

3. EVALUATION IN DISTANCE EDUCATION: EXPERIENCE AT THE UKOU

In this section we give a brief summary of the development of evaluation strategies at the UKOU in order to exemplify some of the important issues.

Three features of the work that are particularly relevant to the preceding discussion are the role of new technology, the question of who best conducts evaluations, and the importance of students' perceptions of their experience of study. New technology can play a particularly important part in the design of an evaluation system for a distance teaching institution. The UKOU has from the outset had a fully computerized student record system, which holds all the data on students' age, educational qualifications and job on entry, and their continuous assessment and overall course results throughout their studies with the university. Not only does this database allow the tracking of individual students' progress, it also enables many interesting links to be traced, for example between the demographics of course populations, assessment scores and students' perceptions of courses. Some caution is required, however, to ensure that the availability of such a system does not lead evaluation to focus solely on the quantitative, and to balance institutional research using computer-based records with other more qualitative approaches.

At the UKOU, the staff of the Institute of Educational Technology play a key role in the design and conduct of evaluations. As members of a separate institute, they can be sufficiently independent of the Faculties to play an impartial role in the evaluation of courses. They can build up and share expertise in the technical conduct of evaluations, and for example develop experience of questionnaire design and the interpretation of results. Also, as course team members they can be constructive as to how course team authors might respond to findings.

Adult students react to courses from a range of perspectives. In particular, as well as their success in learning in the ways intended, they will have views about whether they enjoyed studying and about how the course has changed such affective factors as their confidence or motivation to study in the future. The OU has always stressed the importance of finding out from students how they used the different course materials and their reactions to them. Students are routinely asked to rate the various components in terms of interest, difficulty, helpfulness and usefulness, and to provide information on the time they spend on different aspects of the course.

Perhaps it was because when the UKOU was set up it was so radically different from the conventional universities of the time that so much emphasis was placed initially on monitoring its offerings. Even in 1966, the White Paper on *The University of the Air* stated

> From the outset, it must be made clear that there can be no question of offering to students a makeshift product inferior in quality to other universities. That would defeat its own purpose, as its status will be determined by the quality of its teaching.
>
> (HMSO, 1966, para. 4)

This focus on procedures for assessing and monitoring the quality of

UKOU courses has been maintained and developed ever since the institution's inception. Following Scriven's formative/summative distinction, evaluation is used both at the course design stage and to help assess how well the finished instruction works. It has been the norm not only for new courses to be evaluated from a variety of different points of view, but also for courses to have mechanisms built into their development process for improving instructional materials in the light of feedback received (see e.g. Kaye, 1973; McCormick, 1976; Field, 1982). A wide variety of approaches have been used as detailed in Henderson *et al.* (1977, 1980), Kirkup (1981), Nathenson *et al.* (1981), and Scanlon (1981). Evaluation may be carried out at different times during the life of a course, in a variety of different ways to meet a variety of different needs. As discussed in the last section, one of the key issues is the purpose for which an evaluation is carried out. In the UKOU data are needed for a variety of purposes. Course teams require information about many things, for example, who might take a planned course, how a teaching approach has worked out in practice, how the final course is perceived by the students who took it, whether any parts of the course were particularly difficult or interesting, whether the use of a particular medium is as successful as in other courses, and so on. Most evaluation required by course teams is formative in nature – at least potentially. For even if it is collected after the course is in presentation, teams will wish to act on feedback to improve the course. It is also likely to be fairly detailed, to help in designing improvements, and often the timing of feedback collection is particularly important if great detail is required. In contrast, institutional requirements are for different types of information: for example, management requires comparative data across courses, years and levels of study to check whether quality assurance procedures are working. This is an issue we will take up again in Chapter 15.

Figure 14.1 is a list of methods of evaluation adapted from Kirkwood (1992). We will use it to organize our description of UKOU practice, describing each of the main categories. In Section 4, more detailed case studies will show how the various types of evaluation can interact.

3.1. Types of formative evaluation

Before a decision is made to proceed with a course on a particular topic, market surveys may be conducted to establish the potential student audience, their educational background and which other courses they have taken. Sometimes, once the decision is taken to mount a course, further research is done to investigate the content of similar courses or students' expectations of what the course will provide for them.

As discussed in Chapter 12, while courses are in preparation at the UKOU the draft materials are subject to a large amount of critical commenting. This can be from course team colleagues, from part-time tutors invited to join a readers' panel, or from experts in the area acting as external assessors on the course. These external assessors are formally responsible for monitoring the quality of teaching materials, and must submit a report to the Vice-Chancellor via the Dean of the Faculty before the course is approved.

Developmental testing often takes place during the course preparation phase and involves a group of learners studying a version of the whole course or part of the course and providing feedback on which necessary revisions to the course material can be based. This procedure can be highly formal, whereby the whole course including the assessment material is tried out and credit awarded to successful students (so called 'developmental testing for credit'), or quite informal whereby small pieces of teaching are tried out by students for money or even for nothing! Experience of this procedure is that it is of benefit to both course writers and the students who study the improved version of the course (see e.g. Henderson *et al.*, 1977, 1980).

3.2. Types of summative evaluation

Summative evaluation provides information on courses in use. At the UKOU all new courses are monitored at the end of their first presentation by means of

Formative evaluation

• market research studies

• critical commenting (by experts on materials in draft)

• developmental testing (by students)

Summative evaluation

• student feedback

• tutor feedback

• 'monitoring' of script marked assignments

• statistical item analysis of computer marked assignments

• cross-sectional studies

• developmental studies

• special projects

Figure 14.1. A list of evaluation methods (after Kirkwood, 1992).

an Annual Survey of New Courses. An extract from one of these surveys is given as Figure 14.2. Surveying is a complex and expensive business, as all the data are held in a central database, so that comparisons can be made over different courses and years. As student populations on courses can change over the years and course teams may alter the material in the light of feedback collected during the first year, it is desirable that this monitoring is done several times during the life of the course.

Students are invited to comment on their experiences of studying a particular course by means of questionnaires and interviews. These may focus on the

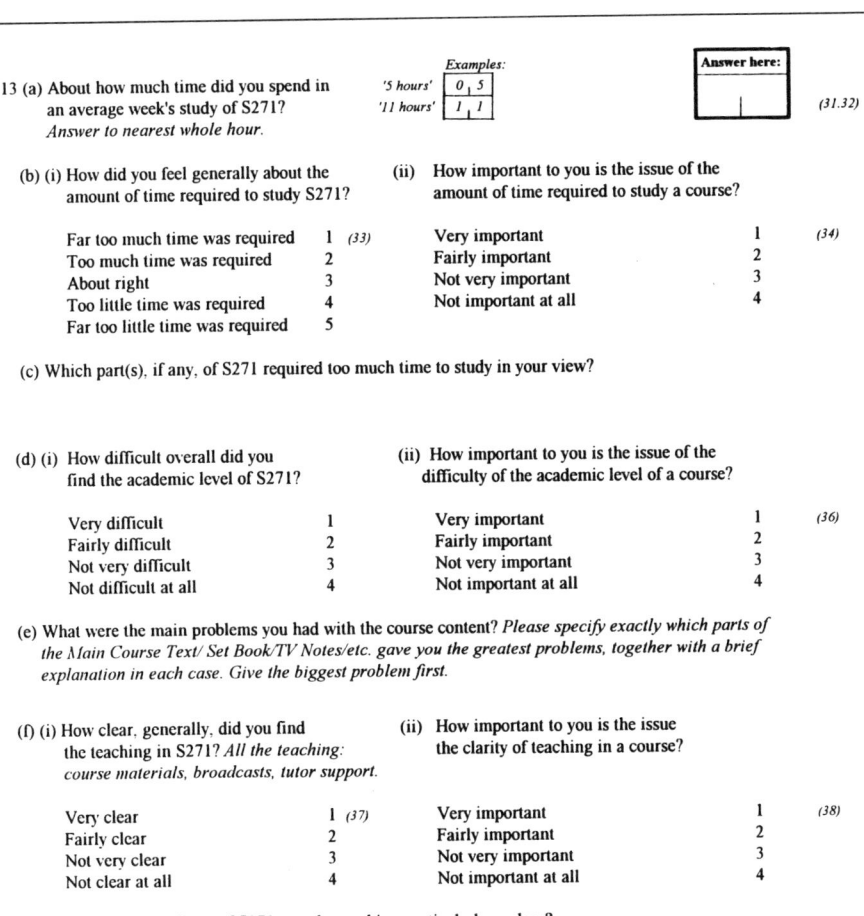

Figure 14.2. An extract from an Annual Survey of New Course survey by which all new UKOU courses are monitored at the end of their first presentation year. The survey contains a number of questions that are common to all courses and a section in which course specific questions are included.

course as a whole, or on individual sections of texts or other components. Questionnaires can cover specific content issues, the general style of presentation, an overall view of the teaching, extent of use of particular media, etc. Questionnaire design can put constraints on student comments, but this difficulty can in part be overcome by including space for open-ended comments. Interviews, either face-to-face or by telephone, can be used to probe issues more deeply, but they are more time consuming to conduct and to analyse.

Information on student performance is available from their results on assessment material; and procedures also exist for monitoring the quality of the assessment material, in addition to the formative commenting described earlier. At the UKOU, all tutors' script marking is monitored, both for grades awarded and for the quality of the correspondence tuition provided. All computer-marked questions set in assignments and examinations have a procedure of item analysis applied to them, to identify questions that have been problematic, those that have performed well in assessing students' understanding and those that have discriminated well. In this context the following comment from Calder is pertinent.

> The importance of student record statistics in providing a statistical context for both the feedback studies and the in-depth studies should not be forgotten. Student drop out, completion of assignments, exam pass rates – all provide comparative background information against which student and tutor feedback can be interpreted.
>
> (Calder, 1994, p. 201)

Sometimes it is necessary to conduct an evaluation of a number of courses with some feature in common, to draw out generalizations about a particular aspect of teaching or use of media – a cross-sectional study. Sometimes what is required is a developmental study – i.e. tracking the performance of students through their study of several courses towards their degree (see e.g. Ross and Scanlon, 1989).

4. EVALUATION FOR SCIENCE STUDENTS

In this section we take two UKOU case studies and explain in more detail how the types of evaluation described earlier have been used in developing science teaching. The first case study illustrates an evaluation designed to investigate student use of a particular medium – computer-assisted learning in science. The second case study illustrates the range of evaluation efforts that were usefully applied in the development of a core course in physics.

4.1. Case study: an in-depth study of use of one part of the media mix

This study (Jones and O'Shea, 1982) was conducted to investigate the very low use of tutorial CAL packages in a particular course on the biological bases of behaviour. The study proceeded in three stages. At the first stage a questionnaire study found that many students never used the optional CAL packages available at study centres which were intended to provide remedial assistance. The reason for this was unclear, so at the second stage an interview study was conducted at summer schools. This interview study established three types of fear which students had about using the CAL package:

- fears about how they would appear to other students or that they might look stupid;
- fear that they might damage the hardware or software;
- fear that their scores on the computer-marked assignments on the course might be compared with their performance on remedial tutorials, to their detriment.

The events that had generated these fears were described as 'bad computer experiences'. Some of these events when described to other students even generated 'second hand' bad computer experiences (Scanlon *et al.*, 1982). The third stage was to echo these fears back to 2,000 students on a range of science courses which offered similar arrangements, in a questionnaire which established that the three fears were not uncommon. An unexpected result was the popularity of using study centres where there was only one terminal because of the privacy. This case study illustrates the useful findings that can come from applying a mix of qualitative and quantitative methods; it is important to note that it was necessary to consider an alternative approach (involving informal interviewing at summer school of students who did not use the CAL) in order to gain some clues about the reasons for low usage prior to collecting numerical data at the final stage.

This case study emphasizes the first of each element of the types of evaluation which were presented as the dichotomised pairs described in Section 2. The style of this particular evaluation was illuminative and exploratory. It was also ideographic, focusing on the description of what actually happened in relation to student use of the medium in a particular context. It was summative in the sense that the course already existed, and its primary purpose was to discover whether the aims of the course team in respect of the use of media had been achieved. But it was also formative in the sense that the result of this evaluation informed the development of an appropriate policy for future uses of CAL in UKOU science courses. It is often possible for data to be used for both summative and formative purposes.

This case study also illustrates the need to mix styles of evaluation to obtain a meaningful picture. Other examples of this mix of methods include the valuable role that snapshot studies can play in gauging the scale of the task in conducting larger pieces of evaluative work, and the eventual need for quantitative work to assess the scale of a problem that may initially have been identified by qualitative pilot studies.

4.2 Case study: evaluation through two generations of a course

The UKOU course *Discovering Physics* has been alluded to in several previous chapters, and provides an interesting example of a course that has been the subject of the full gamut of evaluation procedures from initial conception through more than a decade of presentation to a major rewrite.

Devised in the late 1970s, the course represented a new concept in OU science courses – a 'gateway' course designed to appeal both to physics specialists and to students majoring in other areas. The first requirement for information was therefore a market survey. A draft course description circulated to students then working on a science, maths or technology Foundation course, together with a form on which they could register their degree of interest in the proposals for *Discovering Physics*, suggested that the course would indeed be attractive to students with a broad range of backgrounds and aspirations. Demographic information collected subsequently showed that these forecasts regarding the heterogeneous composition of the student population were in fact fairly accurate, and a special survey (Ross and Scanlon, 1989) later established the extent to which students made use of the 'gateway' nature of this course to tailor their degree profiles in a wide variety of individual ways.

During the initial drafting stages, the course team gave careful consideration to evaluation reports on various innovative teaching strategies adopted in other courses. For example, some time was spent examining student reaction to audiovisual presentations in the science and maths Foundation courses, and an approach was eventually adopted that built on both the maths use of audio commentary to guide students through conceptually difficult material presented on hand-drawn frames and the science experience of using audiocassettes to discuss diagrams. The resulting audiovisual sequences proved so successful, and were on the whole so well received by students, that a very similar style was adopted both in the next physics course to be written (see Figure 7.5) and in the next generation of the science Foundation course.

When the course material was in good draft form, thirty students, with a range of educational backgrounds, were recruited to developmentally test the course for credit, the course fee being waived in return for their evaluative responses. Comments were collected regularly, by means of feedback questions inserted at the end of virtually every section of the main texts. An example of the level of detail of these questions is shown in Figure 14.3. As a result of this feedback, many minor and several major changes were made before the first full presentation, including the complete removal of about one-eighth of the original material and the incorporation of a new, mid-course revision section with a specially made TV programme to accompany it.

The course proved continuously popular, with about 700 students each year through the 1980s. In its first year of full presentation, student reactions were evaluated by means of feedback blocks on computer-marked assignments and questionnaires; whenever possible, further small changes were made to the material to address perceived difficulties. Student performance was routinely monitored by means of statistical item analysis of continuous assessment and examination performance. As a preparation for the rewrite, a national meeting of tutors was convened after the seventh year of presentation, and a new round of student feedback solicited during the eighth. The tutor comments were particularly influential in focusing the course team's attention on the need for greater prominence to be given to problem-solving. This, together with the evaluation studies of the problem-solving element in a number of mathematics courses eventually led to the protocol already discussed in Figure 5.8. The tutor feedback also helped to pinpoint the concepts that caused students the most difficulty and suggested areas in which cuts could be made; feedback from students was more useful in highlighting ideas or examples that had

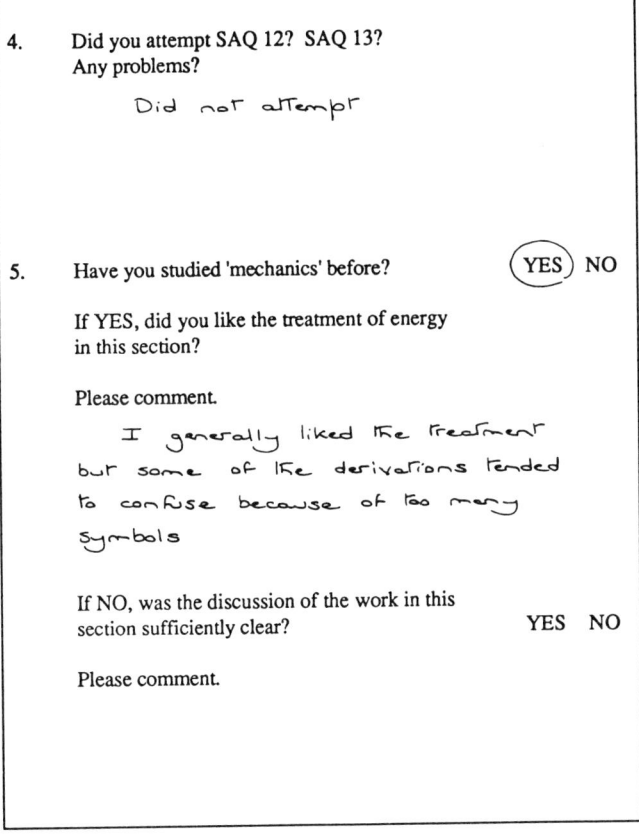

Figure 14.3. An example of the kind of detailed feedback questions inserted at the end of each section of each unit during the developmental testing of the original version of the UKOU course *Discovering Physics*.

proved particularly interesting or motivational, in quantifying workload, and in showing the relative popularity and effectiveness of various elements of the media mix.

During the rewriting phase, reactions to the proposed problem-solving protocol, to associated new formats such as worked examples (see Figure 12.6) and to a different way of starting the course were again obtained by developmental testing – by unpaid student volunteers – of drafts of the first four units. The complete new version of the course was first presented in 1994, and a new round of evaluation immediately started (Ross and Scanlon, 1995).

The style of evaluation adopted in this case is more formal, more statistical and less exploratory than that outlined in the previous example, but again illustrates the benefits of an approach that combines information from a number of different sources. As noted by Harris and Bell (1986), the element of triangulation in evaluation studies is very important. Evaluation that looks at the educational effectiveness of whole courses is also clearly linked to quality assessment and assurance procedures.

5. QUALITY ASSESSMENT, QUALITY ASSURANCE AND DISTANCE EDUCATION

The most recent developments in evaluation are in the area of 'quality' assessment and assurance. Although there are many sightings of the word quality in recent writing, there are few definitions of what it is supposed to mean in an educational setting. The notion of quality was imported by education managers from the 'total quality management' approach currently popular with management consultants. In its original manifestation it was applied to the manufacturing industry, where quality was defined as 'fitness for purpose' but this does not translate well into the educational sector. Nevertheless, governments are increasingly imposing the idea on educational institutions, with money attached to success. In Scotland, where the first teaching quality assessment exercise of its kind in Europe was conducted, institutions achieving 'excellence' were promised a 5 per cent increase in funded student places for the next five years. These institutions were assessed on a quality framework consisting of curriculum, environment and resources, teaching and assessment, student guidance and support, and outcomes and quality control procedures. New frameworks are currently being established throughout the UK, and similar trends are evident in other parts of the world. For example, Australia's universities are introducing a federal government funded quality enhancement scheme which allows for additional funds of up to 3 per cent of an institution's operating budget to be awarded on the basis of an audit. The following quotation refers to the traditional view about how quality is maintained in British higher education.

The quality of undergraduate education in Britain has traditionally been defined in terms of a special intimacy between teachers and students, which is thought to depend on small group teaching, which in turn requires generous staff–student ratios. Neither of the latter two stages in this supposed syllogism necessarily follows.

(Scott, 1994, p. 53)

Current debate in distance education (see for example Atkinson *et al.*, 1991; Barnett, 1992; Nunan and Calvert, 1992) focuses on such issues as:

- the extent to which the learner can or should be involved in the assessment of quality;
- the importance of 'added value';
- the desirability of instituting 'management for quality' rather than 'management of quality';
- the current climate of rapid educational changes, in which institutions and teaching practices need not only to be fit for their purpose in the short term but adaptable and innovative in responding to external change in ways that maintain, or better still enhance, quality.

In one sense, distance teaching institutions are better set up for quality audits than their conventional counterparts, since they have usually incorporated evaluation mechanisms from the outset and tend to have extensive computerized databases of student information. It is also comparatively easy to design evaluation instruments for course components such as texts or videocassettes. However, the distance teaching institutions have a much bigger problem in evaluating the 'services' aspect of their provision, such as the tutoring, counselling or responses to administrative queries. While these aspects are included in UKOU evaluation, in an attempt to identify the sources of difficulties that students experience with such services, Robinson (1994b) points out that it is more difficult to introduce suitable quality assurance procedures in relation to services than in respect to course material.

6. CONCLUSION – PROBLEMS AND POTENTIAL OF EVALUATION

Evaluation activities can be particularly helpful in highlighting problem areas in courses, but they do not of themselves lead to improvements in related materials. For that, solutions need to be found to the problems. Often, once action towards making suggested improvements is begun, the changes can become more extensive than is strictly necessary. Melton and Scanlon, writing about the evaluation conducted for the rewrite of the UKOU science Foundation course in the early 1980s, noted that

Evaluation may expose weaknesses and, under particular conditions, may identify possible solutions, but it does not in itself determine change. This depends on a variety of other factors such as the perceived relevance, timeliness and helpfulness of the data, on whether or not authors accept the data and related findings, on their ability to find solutions to problems identified, on the quality and quantity of advice available to them from coauthors and evaluators, on whether or not they accept such advice, and on whether or not they are able to transform such advice into related improvements.

(Melton and Scanlon, 1982, p. 71)

Where does this ability to transform advice into improvements come from? How can one best collect evaluation data that can be meaningfully used by authors? One way forward is to use the course team process as a learning tool. The most useful thing that evaluators can do is sometimes to write teaching material themselves, and the most useful thing that authors can do is sometimes to engage in data collection, so as to improve their understanding of the actuality of student learning.

Another purpose that evaluation can serve is that of staff training. For example, the distance education centre of the University of South Australia tries to involve individual teachers in course feedback because they see it as a necessary part of staff development (Nunan, 1992). Teachers are encouraged to conduct their own evaluations, modifying for their own purposes templates of questionnaires on a range of questions dealing with such areas as subject-matter, resources provided, teaching and assessment arrangements.

It is part of the new orthodoxy in higher education that the effectiveness of teaching needs to be assessed, and indeed many conventional institutions now run evaluations of courses by students. It might seem churlish to question this practice, but it can lead to problems. For example, a recent report on equal opportunities policies (Bagilhole, 1992), mentions problems with racist and sexist comments in student evaluations. Also, response rates are often low and therefore the results statistically unreliable.

We have mentioned the link between evaluation and quality assessment and assurance a number of times. The danger here lies in the possibility of overemphasizing quantitative measurements at the expense of other approaches to the measurement of quality. Quality in education should mean something more than 'fitness for purpose intended'. The last thirty years have seen the development and application of various evaluation methodologies. The challenge now is to apply some of the lessons learned about the conduct of meaningful evaluation to assessment of quality in the educational arena.

PART C:

Future Prospects

15

TECHNOLOGY AND PEDAGOGY

We want to gain insights into what it means to share knowledge; how knowledge can be captured and conveyed; how to exploit new technologies to satisfy the ever-increasing appetite for communication bandwidth; how to meet the needs of mobile students; what new methods will assist disabled students; how to work in groups with students scattered around the globe; how to harness software in the service of human understanding.

(Eisentadt, 1995, p. vi)

1. INTRODUCTION

In this book we have established how open and distance learning in science can be achieved. We have discussed the particular benefits of multiple media, distance teaching approaches for science education and illustrated how some of the problems relating to distance education in science can be triumphantly overcome. In this chapter we will recap a little on some of these themes; we also will allow ourselves the luxury of speculating, from the perspective of current knowledge, on the new challenges that may soon face distance educators, on where new ideas in science education are heading and on the problems, solutions and stimulation for instructors provided by new technologies.

In Chapter 2 of this book we highlighted some key challenges for science distance educators, including the problems for course designers of dealing with open access, or open entry policies, and free choice in degree profiles, and the problems of providing appropriate practical experiences for home-based learners. We have illustrated the variety of ways that this practical experience can be provided using a variety of media, and in this chapter we will consider how new technological possibilities can develop this approach further.

The audience for science knowledge is ever growing. In England and Wales the advent of the National Curriculum in schools has led to an increasing population of students who have some science background and are in a position to consider further studies in science. This new emphasis on science is in part a response to the results of international studies of science achievement (see e.g. Postlethwaite and Wiley, 1992) which show that other countries often achieve larger participation rates in science. Another feature highlighted by these studies is the low position that Britain occupies on the international scale in relation to the participation of women in science. The UKOU has been praised for its contribution towards increasing the participation of women in higher education generally, and this is a feature of open and distance education, especially in scientific and technological subjects that may particularly commend itself to other countries with similar imbalances to address.

We described in Chapter 2 the 'science for all' movement, which focuses on alternatives to the science curriculum with the specific aim of increasing participation and understanding of science at all levels (Fensham, 1993). The USA has set itself the goal that *all* students become scientifically literate by the end of their twelve years of compulsory schooling. In pursuit of such goals, the National Science Foundation inaugurated a programme in 1991 to support a selected group of states in their reform of the science education in their schools. Significantly, the proposal solicitation (National Science Foundation, 1990) 'stressed the need to involve state leaders, school system leaders, *leaders in science-rich institutions (universities and colleges*, museums, science laboratories)' (Raizen, 1993, p. 51). Our italics show the importance of higher education in helping to reach this goal, but also the changed perception of the role of such non formal settings as museums in teaching and learning science.

The following quotation illustrates the role that non formal science education can play in motivating people towards the study of science:

> Lord Snow and his 'two cultures' did much damage to public science by shoving it into a limbo of its own, detached from the mainstream culture of the humanities. It is slowly being dragged back by talented publicists such as Hawking, Wolpert, Feynman and Dawkins. Popular science books sell well to adults, as popular electronics sells well to children. They do so not for their relevance but for their irrelevance, for taking us on a voyage we never knew existed to the farthest reaches of the imagination. That is exciting science.
>
> (Jenkins, 1995, p. 29)

Effective science teaching will capitalize on this sense of excitement and exploration, but the open and distance education institutions have to go beyond the level of popular science books.

> The science that is needed by an advanced industrial society cannot be learned by watching mother, sitting next to Nelly, watching *Tomorrow's World* or *Horizon* on the TV, reading the newspapers, poring over teach yourself books in the evening, or even by apprenticeship to a practical craft. Our technological civilisation . . . would slowly collapse if tens or hundreds of thousands of people were not spending some of the formative years of their lives learning science systematically from professional teachers.
>
> (Ziman, 1986, p. 132)

We hope to have shown in this book how multiple media instructional methods can contribute to such formal and systematic learning, while still conveying some of the wonder and excitement of scientific discovery. One of the challenges for instructors is to maintain, or even enhance, both these strands of the educational experience as new technologies offer new pedagogies.

2. CONVERGENCE OF CONVENTIONAL AND DISTANCE EDUCATION SYSTEMS IN THE UK

In the last ten years the participation rate, i.e. the proportion of eighteen-year-olds, entering UK higher education has risen to 50 per cent, in fact reaching in advance the target set by the government for the year 2000. However impressive this expansion is, it is less striking than the trend towards part-time study which in the same period has increased even more (O'Shea, in press), both through the UKOU and through conventional higher education institutions.

It is UK government policy to increase participation in undergraduate education in mathematics, science and technology, whether full time or part time and whether by distance or conventional methods. There is also pressure on tertiary institutions to increase their income from courses of high perceived value offering specialist qualifications such as an MBA or MSc. One way to achieve this is by special versions of courses already developed for overseas markets, using distance education methods.

Is this convergence in the types of students being taught by the conventional and distance education sector matched by any signs of convergence in methods? Two examples suggest that it may be. One is the Open Learning Foundation (OLF) – a consortium of new universities formed to share the cost of development of distance education material in core courses. The other is the Teaching and Learning Technology Project (TLTP) initiative set up to facilitate efficiency gains in introductory and service courses across the higher education sector, and to find new ways of coping with increasing numbers of students by producing technology-based teaching which is usable in a number of institutions. The TLTP is 'the largest ever earmarked injection of public funds into learning technology in higher education' (Darby, 1993, p. 2). Forty-three collaborative projects were funded from August 1992 (from a budget of £22.5 million available over three years). 'Multi institution consortia form the backbone of TLTP' (*ibid.*, p. 27) and entail courseware development involving several academics. Although the TLTP programme has not been without criticism in terms of its objectives, and its technical base, it seems likely that it heralds the way forwards for developments in conventional higher education, especially in relation to introductory and service courses where there is a large potential overlap in the content of courses from different institutions. Both the OLF and TLTP initiatives are important examples of the need for collaboration in ventures involving the design of distance education materials.

Economies of scale are also important in the distance education sector. Most UK universities are too small to mount local distance education operations, but there are some interesting examples of foreign operations such as Wye College's part-time degree offered in more than 100 countries (Teshome, 1994).

3. BEYOND CONVENTIONAL DISTANCE LEARNING AND EDUCATIONAL TECHNOLOGY

Evans and Nation (1989) have labelled the UKOU as 'instructional industrialism' due to the production of teaching texts by complex divisions of labour in which there are parallels with mass production. The challenge is to adapt this model of distance learning to apply in the new era of communications technologies where the 'car production' metaphor is less helpful. A variety of educational technology metaphors are in use at the UKOU, and Harris (1987) has criticized various aspects of the instrumental approach of the institution's early years. However he also offered two hopes for the future. The first relates to the enormous influence that educational technology pioneered at the UKOU has had on educational practice: 'Educational technology was able to expose so many of the assumptions of conventional practice, and perform really useful work in testing the implicit claims to effectiveness of conventional teaching' (Harris, 1987, p. 63). The second relates to the suggestion that distance teaching systems can engender in students more independence of thought than some other forms of education:

> Educational knowledge itself can radicalize and raise the critical consciousness of its receivers. Although there are grounds for pessimism about the possibility as a major outcome, it would be unwise to dismiss it completely. On the whole, students might use their freedom from the system at a distance to pursue cynical strategies of 'playing the game' or opt for 'surface level processing', as the studies showed. But some students do gain considerable personal insight, self confidence and ability to reflect upon their own surroundings.
>
> (Harris, 1987, p. 149)

In Chapter 3, we discussed social constructivism as the predominant view of educators seeking to understand the processes of learning science. Related to this view is an attempt to apply a postmodern critique to distance education. Bergquist (1993) describes postmodern organizations having features of premodern oral cultures and notes that modern literate cultures are defined by electronic documents. Applying this notion to education, Davis (1994) suggests that learning organizations need to think about how changes in media have altered the meaning of the documents used in distance education systems. He believes that there is now a need for documents which are of two rather different types. Stable long term documents establish the system; more conversational communications (whether actual conversations or e-mail conversations) may be more effective in allowing change or learning to take place.

One of the questions we will investigate in this chapter is what implications developments in technology have for the students' experience of distance learning. There are a number of sub-questions related to this general question, among which are:

- How does new technology challenge the primacy of the basic teaching text?
- How do new methods of communication alter the dialogue between students and between students and tutors?

One consequence of the introduction of electronic communication is that the notion of 'tutorial in print' decribed in Chapter 6 becomes transformed to 'dialogue in the ether'. The consequent changes in pedagogy are the subject of the next section.

4. CONVERGENCE OF COMMUNICATION AND COMPUTER TECHNOLOGIES FOR TEACHING AND LEARNING

There have been a number of experiments in recent years that have attempted to use the new possibilities offered by technology for teaching at a distance. Some of the established UKOU developments were described in Chapters 8 and 9. Advanced learning technologies are quickly becoming established in many areas of higher education (see e.g. Mason, 1989a,b; Mason and Kaye, 1989; McConnell, 1992).

However, until now the technologies for distance learning have been seen as essentially complementary, if not almost interchangeable. An intermediate stage was the advent of multimedia where, as we described in Chapter 9, the integrated use of media offered new potential in the design of specific pieces of teaching. Now, with the convergence of computer and communications technology, there are new challenges and opportunities. For the first time it is becoming possible to do more than imagine the shape of a teaching system in which all the functions of the different media, including communication and administration, are available for the distance student at home in the form of a personal workstation and communication system. O'Shea has outlined his vision of what he calls 'the virtual university' of the future, in which all the functions of a distance teaching system are facilitated using new technology.

> In the Virtual University of the future students will work with each other, with tutors and with remote interactive educational media. They will be able to interact synchronously, for example, by screensharing and asynchronously, for example using bulletin boards. They will not need to work from a single physical location and they will access all the services – fee payment, transcript production, course selection and learning support

– via their electronic address. The technology will support their learning in various ways via a range of virtual facilities including simulated laboratory equipment and remote library services. The single most important facility for the student will be virtual electronic notebooks which will serve as study organisers, memory prosthetics and interactive media in which they will manipulate and store symbols, images, text and selected multimedia interactions.

(O'Shea, in press)

His belief in the viability of this concept is based on the fact that a number of the necessary constituent parts of the system for such a virtual university have been tested out at the UKOU and elsewhere: teaching by e-mail, computer conferencing, aspects of the tutoring and counselling role, assessment and even collaborative production of course materials made available by communications networks between authors in a number of countries. Several attempts have been made to teach students via e-mail (see e.g. Veasey D'Souza, 1991; Wild and Winniford, 1993) and other distance learning institutions, as well as the UKOU, have also become involved in such systems. For example, Ellerman *et al.* (1992) developed a system called StudieNet that allowed Dutch Open University students to share information, talk to their tutor, and submit assignments using e-mail. McConnell (1994) reports on the experience of staff and students on a two-year part-time computer mediated MA in Management Learning offered by Lancaster University, taught by five residential workshops with substantial periods of on-line group work involving four or five participants:

A major concern on the programme is with the social aspects of learning via CSCL media. Our starting point is not the technology; we see that as a tool, a means to an end. Rather we are concerned to establish and maintain a supportive learning environment where participants and staff are involved in actively constructing meaningful learning relationships that are acceptable to them all. We use the term learning community to emphasise the social and learning relationships that are encouraged online and in the workshops. The programme is concerned as much with participants' self and professional development as with their academic learning.

(McConnell, 1994, p. 61)

The students he quotes have a very positive view of the learning experience they engaged in but see the course as something which has completely taken over their lives in terms of time. This is in line with other findings by Axelrod (1990), McGrath (1990) and Pinsonneault and Kraemer (1990) who all comment on the structuring implications of what McConnell calls the 'fluidity of time in electronic environments'. In the longer quotation above, McConnell highlights another advantage of the Virtual University for distance learning students – the possibility of benefiting from collaborative learning. Two other experiments that are addressing this issue are the CO-LEARN European research project, and the JITOL project. The CO-LEARN project is a series of experiments using IDSN (Integated Services Digital Network), audiochannels and multimedia and exploring a number of modes for groupworking (see Derycke and Kaye, 1993). The JITOL (just in time open learning) consortium is exploring self directed open learning at a distance using information technology, and is operating experiments in three areas: training of trainers and teachers in advanced learning technologies, training health professionals, and training people working in banks. We will return to these topics later in the chapter, but first it is appropriate to discuss in a little more detail the nature of the technological developments which are currently having such an impact on the educational world.

New technologies for communication

In the last few years a number of technologies which have the potential to facilitate learning have become more widely available. Developments related to telecommunications in particular have great relevance for distance learning and training.

The Internet is an evolved version of the Advanced Research Projects Agency of the Department of Defence high bandwidth network (ARPAnet) which was set up in the mid-1970s to support electronic communication between research laboratories (mostly based in universities with ARPA contracts). This network has now been extended to most academics worldwide.

> Once you are connected to the Internet, you have instant access to an almost indescribable wealth of information . . . Through electronic mail and bulletin boards . . . you can use a different kind of resource: a worldwide supply of knowledgeable people, some of whom are sure to share your interest, no matter how obscure.
>
> (Krol, 1994, p. xix)

The World Wide Web is a set of conventions to make available local information on file servers to people using the Internet to gain access. The growth of this Web is interesting in terms both of the number of new file servers making information available to remote users and the number of users using these facilities to roam 'the information superhighway'. The scale of science resources available on the Internet via the World Wide Web is truly impressive. It is possible to participate in experiments in a number of continents via the JASON project – for example, to collect and identify spiders and feed back results to a woman scientist in the USA who is seen giving a brief 'talk' on her research project. It is possible to 'visit' exhibits at the Exploratorium in San Francisco, or to conduct the dissection of a frog via a multimedia simulated dissection at the University of Virginia School of Education.

Eisentadt has asserted that the UKOU (and presumably other distance education institutions too) ultimately will develop complete courses using the Web, but points out that there are many things still to consider.

World Wide Web: we are developing the first complete profile of courses which can be taken by students from anywhere reachable via the Internet, and which lead to a batchelor of arts degree recognised by the OU and others. This activity will require the development of new styles of tuition, course presentation, delivery and facilities management.

(Eisenstadt, 1995, p. vi)

Not least among the questions raised by these new styles of course development is the issue of how to finance such ventures. With so much high quality material available free in the system, it seems clear that institutions wishing to charge for such courses will be charging for accreditation rather than for supply of materials.

The possibility of providing access on line to the information available via the Web suggests that a new area in which pedagogy needs to be developed is how to properly support the development of resource-based learning.

Of particular relevance to distance education and training are a number of collaborative experiments involving the use of telematics that have been conducted recently. The basic requirements are a workstation consisting of a personal computer, a modem telephone line and appropriate communications software for the individual student and a main computer where the group software is held and which allows access to other networks such as the Internet. New methods other than conventional telephone lines are being developed to allow video conferencing and transmission of complex multimedia programs, including direct broadcast by satellite, fibre optics, and a digital telephone service. The UKOU was the lead partner in the European funded JANUS project (Joint Academic Network Using Satellites) which aimed to build a prototype satellite-based telematic network linking six sites (UKOU, Dutch OU/European Association of Distance Teaching University headquarters, Portugal Telecom, Athens, Greece, Chania, Crete and Lahti, Finland) and to deliver courses over the network. One example of such a course was *What is Europe?*, an EADTU course presented across Europe by the UKOU with tutorial support via JANUS. Another recent development is the Integrated Services Digital Network (ISDN), with digital telephone lines; this allows faster transmission of data which is particularly important for videoconferencing. In the next section we will examine recent UKOU experiments which have used these new technologies to support collaboration of various types.

5. UKOU EXPERIMENTS IN COLLABORATIVE USE OF NEW TECHNOLOGIES

One key benefit for distance learners of these technological developments is the opportunity for students to collaborate together in their learning. Until recently the opportunities for the distance teaching institutions to make full use of remote group work seemed far in the future. Despite critiques such as those of Evans and Nation (1989), and the success of conferencing experiments, the opportunities for extensive collaboration at a distance were not available. Such projects as were launched were relatively modest in scope: the Modems in the North (MINT) project spread access to modems in the Highlands and Islands of Scotland, while the Shared Whiteboard project (Weston, 1993) tested out a British Telecom technology in the context of second level UKOU science tutorials spread over a number of study centres. However several recent projects have broken new ground.

Fourteen hundred graduates are currently (mid-1995) halfway through their UKOU 18 month postgraduate teacher training course. They are supplied with Apple Macintosh computers and modems; they are able to talk to their tutors through electronic mail, join electronic conferences on their subject area and access the UKOU information system, using a conferencing/mailing system called FirstClass which has the added benefit (as does the latest version of CoSy, the conferencing system described in Chapter 8) of a graphical user interface for the conferencing system using Windows, Icons, Mice and Pulldown menus. This system will be used for other UKOU courses from 1996. Mason (1995) reports that in 1995, UKOU students on the *Fundamentals of Computing* course are being offered the 'possibility of participating in structured activities and discussions through combining the facilities of First Class and the World Wide Web' (p. 7). Additionally, 1000 students on the remake of the *Introduction to Information Technology* course will have access to the Internet and World Wide Web; conferencing will be centred round a collaborative project, to be submitted electronically. Course materials from a postgraduate course on LISP programming are currently offered via the Internet and it is possible to download versions of the course material. This development follows an experiment in which the course text materials were supported through conferencing managed by two of the academics on the course team, and is one of 'a number of small population courses taught totally electronically either through conferencing or over the Internet' (Greenberg, 1995, p. vii).

Renewable Energy Technology was an experimental course presented to students from Britain, Australia and Finland (Alexander and Mason, 1994). Students were provided with text, video and an Integrated Learning Support Environment (ILSE) on their home computers. This ILSE, using a room metaphor to describe it, consisted of a meeting room (using FirstClass as an asynchronous computer conferencing system to link all students and staff together), a library of course materials (using a CD-ROM containing a range of reference materials including an electronic version of the course book, two audiovisual lectures, computer models and one hundred reference articles) and a study for private working (using ClarisWorks which includes word processing, spreadsheets, painting and drawing tools). The main aim of the experiment was to find ways of providing students with meaningful experiences of technologically mediated collaboration to enhance their learning. With this end in view, the course contained three collaborative projects: the production of a joint document based on resources in the CD-ROM Library, problem solving in a virtual world using a spreadsheet model, and a role playing activity. Alexander and Mason (1994, p. 14) were happy that the course 'enabled students to work efficiently, produce much more sophisticated output than using conventional media, and collaborate effectively'.

Another pilot project that may point the way to the future is the *Virtual Summer School* run experimentally in 1994 by a *Cognitive Psychology* course team. This gave some useful information about student reaction to collaborations mediated by technology, where, as well as using asynchronous communication, students were able to participate in videoconferences. A group of students who were unable to attend an essential residential school for family or health reasons were offered the option of using their home computers with modems and custom software, to work in project teams over a three week period. Eisenstadt writes

> The students are able to participate in group discussions, run experiments, obtain one to one tuition, listen to lectures, ask questions, . . . work in project teams, undertake statistical analyses, prepare and submit nicely formatted individual or joint written work, prepare plenary session presentations, and even socialise and chit chat without leaving their homes. The software packages supplied to students emulate many aspects of the residential summer school but without requiring physical attendance.
>
> (Eisenstadt, in press)

The students conducted experimental projects and wrote simple computer programs to illustrate various aspects of cognitive processing. They even 'attended' a guest lecture by hearing and seeing, by virtue of videoconferencing software, an eminent expert speaking in California and participating in the question and answer session which followed. The evaluation of the experience is currently being documented but early reports (Issroff, 1994b) suggest that the use of 'off the shelf' communication products supports remote group work reasonably well, and overall students rated the experience as very successful. One student comments on the benefits, mentioning

1. the possibility of discussing your problems immediately with the tutors is a great relief.
2. the excitement of working in groups and seeing the experiment progress via different points of view and finally coming to the same conclusion.
3. the final assessment of the benefit I derived from this technological innovation is enormous: psychologically and academically.

(Eisenstadt *et al.*, in press, p. 31)

One problem in using these experiments to envision the Virtual University of the future is in imagining how a similar experience could be delivered to larger groups of learners. Eisenstadt himself casts doubts on the possibility of 'scaling up' the experiment to realistic numbers of students because of the difficulty of providing the necessary level of tutorial support.

In recent years a number of researchers have investigated what happens when students work collaboratively. They have demonstrated that in many circumstances students working together on a problem benefit more than those working alone (see e.g. Slavin, 1983). There is less agreement on how these benefits arise, how such effects should be measured and how such groups of students should be constructed for maximum learning gains. For example, some workers attribute learning gains to the increased cognitive resources available for solving the problem (i.e. more people having more ideas), some to increased verbalization while working, some to the beneficial effects of externalizing conflict (i.e. having someone else to point out opposing arguments) and some to a process described as the co-construction of knowledge. So there is still a difference of opinion as to exactly how working with a partner enhances learning, and, in particular, how well adult learners perform in the collaborative setting. Benefits also appear to be dependent on a number of task specific factors. Hence research on collaborative learning raises a number of questions about how such group working can best be facilitated. For benefits of collaboration to appear students must, according to Slavin, be provided with a suitable cognitive task structure and an incentive structure, and appropriate assessments have to be developed. However there is general consensus that the benefits of groupworking are worth striving for.

> Learning collaboratively implies peer exchange, interaction amongst equals, and interchangeability of roles, such that different members of a group or community might take on different roles (learner, teacher, information seeker, resource person, facilitator) at different times depending on needs.
>
> (Kaye, 1991, p. 4)

Distance education has particular problems to overcome in facilitating group work. We have discussed the important role that residential schools, and self help and tutorial sessions can play in helping students to form appropriate groups in which they can reap some of the benefits of collaborative learning. Eisenstadt also raises the need for better understanding of the educational issues involved, including the plea that 'we must obtain a better understanding of the nature of *remote* groupwork' (our emphasis). It is essential that projects be conducted that can give guidance about the ways in which remote collaboration can be best mediated.

SharedARK (Smith, 1992), designed to encourage shared work on a science problem, has in some senses the features of a prototype system for future distance education. In SharedARK a single user site consists of a computer workstation, an audio headset, a video monitor and a videocamera. Each site is physically isolated from the others but connected to them via a computer-switchable audio-video link and high performance computer network. The system supports the possibilities of interaction between other students and tutors over video links and of distributed interactive simulation laboratories. Experience shows that collaboration via computer is quite successful (Smith *et al.*, 1991) and although associated technology is currently too expensive for widespread use, the technologies are now becoming available for such interactions to be designed as part of a distance learning course.

The quality of image provided in the videolink in the SharedARK was superior to the video involved in the Virtual Summer School experiment described above for technical reasons. In the latter case the use of video-communication was not felt to be vital to the development of co-operation between students, while in the SharedARK experiment, features like gaze negotation were found to be a very significant part of the development of successful remote problem solving, although there is some controversy over the precise importance of eye contact (see e.g. Sellen, 1992). The possibility of allowing science students at remote locations to work co-operatively on simulations is very exciting. With the advent of high bandwidth networks described earlier, an increasing number of individuals will be able to use such technology with improved videolinks. For example, it is possible that the Virtual Microscope described in Chapter 9 may be developed to enable students to use it remotely.

A move to supporting collaboration at a distance for a large number of students along the lines of a Virtual Summer School for science is we feel someway in the future. However access to the resources available by the World Wide Web is, as we have described above, somewhat closer.

Other key points raised by new developments are the different ways of resourcing collaboration that it offers, and the role of remote experimentation. It is useful to bear in mind the comments made by McConnell (1994, p. 197) in reviewing the CO-LEARN and JITOL project: 'just having the technology does not in itself give rise to collaborative learning. Learners need to have a context in which to understand the purpose of the collaboration'.

It is essential for course designers to consider the cultural context in which they will be asking students and tutors to collaborate.

6. MEDIATION OF PRACTICAL WORK BY TECHNOLOGY

Provision of practical work is one of the key challenges for distance educators in science. We have argued earlier in this book that one of the reasons for practical work being a required part of science instruction is because it is understood by teachers and students to be typical of the work that many real scientists do. However, developments in information technology have enhanced both the measurement and data handling capacities of computers and the communication facilities afforded by the new networks. Both these changes have significantly affected the way in which scientists carry out the research component of their work. Hiltz (1984) has given an account of an early experiment on the impact of new communication networks using the EIES conferencing system. Most scientists now have a workstation in their office, connected to a university communications network. Many access research data on computer files located on a server in their own or other departments. They use computers to measure and analyse data from experiments. Scientists can guide their analysis by using a variety of tools and can transmit text and diagrams to colleagues in other locations for comments and revision. In the future, support for such activities will be likely to include synchronous communication via videolinks and screen sharing of the kind described in SharedARK.

What implications are there here for the role of computer mediated experimental work for science students? The resources which support scientists in their research work are very expensive, but the resulting networks and information could be made available for educational purposes very cheaply. Allowing students directly to manipulate data associated with new discoveries about geology or climate that they may have heard about via television or other news media only weeks previously, and to use computers comparable in power to the workstations available to many university scientists, should have powerful results in motivating them towards science. Most importantly of all, developments in these kinds of technology are changing the nature of a scientist's work and therefore such changes must be reflected in science instruction.

7. REMAINING CHALLENGES FOR OPEN AND DISTANCE LEARNING IN SCIENCE

An important further challenge for distance educators is to increase access to their courses. There is therefore justifiable concern that the new technological developments may operate as a disincentive to certain categories of student. Although there have been similar arguments about requirements for TV or video ownership, which have now rather faded into the past, the issue of access remains an area where course designers need to be vigilant. UKOU experience certainly suggests that the introduction of new technology can discourage some groups of learners. For example the introduction of a personal computing requirement to the technology foundation course resulted in a drop in recruitment to the course among women. However much progress can be made by targeted advertising and by providing loan pools of equipment.

In the early years of the UKOU, the then Chancellor exhorted the University to be 'open as to people, open as to places, open as to methods and open as to ideas'. We consider that the interaction of technology with the developing pedagogy of distance learning is likely to be very beneficial for science learners, so long as appropriate policies are developed to maintain access.

8. POSTSCRIPT

In this book we have outlined how open learning and distance teaching materials can be applied to the learning of science. We leave the final word to a female science student from the UKOU who, in answer to the question 'How has the OU changed you?' replied:

> I've always been interested in how come? I was the elephant's child for questions never satisfactorily answered. And I believed that if I could just once get there, I would be able to find out *how* to find out the answers to any questions that I wanted to ask. I may choose not to ask it, but at least I am totally certain I can go and explore anything. It's this precious tool of being able to go find out, that I never had. I knew a little bit about how a scientific experiment was conducted, but I'd never actually done one. And now I have. And I know how to do it, I would know how to analyse it, and I would know whether it meant anything or not. The OU has opened up every possibility.
>
> (Lunneborg, 1994, p. 45)

REFERENCES

Alexander, G. and Mason, R. (1994) Innovating at the OU: resource-based collaborative learning, CITE report no. 195, Open University internal report, Milton Keynes.

American Association for the Advancement of Science (1989) *Science for All Americans*, AAAS, Washington, DC.

Aspden, P. (1973) An investigation into the effect of prior laboratory experience on the learning of practical skills, Open University internal report, unpublished, Milton Keynes.

Assiter, A. and Swann, K. (1994) *Using Records of Achievement in Higher Education*, Kogan Page, London.

Atkinson, R., McBeath, C. and Meecham, D. (eds.) (1991) *Quality and Distance Education*, Tenth bi-annual forum of the Australian and South Pacific External Studies Association.

Auchterlonie, A. (1989) A case study in educational broadcasting campus radio, in N. Paine (ed.) *Open Learning in Transition*, Kogan Page, London.

Ausubel, D. P. (1963) *The Psychology of Meaningful Verbal Learning*, Grune and Stratton, New York.

Ausubel, D. P. (1968) *Educational Psychology*, Grune and Stratton, New York.

Axelrod, R. (1990) *The Evolution of Cooperation*, Penguin, Harmondsworth.

Bagilhole, B. (1992) On the inside: equal opportunities in academic life, Loughborough University internal report, Loughborough.

Baird, J.R., Fensham, P., Gunstone, R.F., Penna, C. and White R. T. (1991a) Challenge: a focus for improving teaching and learning, paper presented at the AERA meeting, Chicago, April.

Baird, J. R., Fensham, P.J., Gunstone, R. F. and White, R. T. (1991b) The importance of reflection in improving science teaching and learning, *Journal of Research in Science Teaching*, Vol. 28, pp. 163–182.

Barnett, R. (1992) *Improving Higher Education: Total Quality Care*, SRHE and Open University Press, Milton Keynes.

Bates A. W. (1975) Student use of Open University broadcasting: television, learning and distance education, Institute of Educational Technology papers on broadcasting, no. 44, Open University internal document, Milton Keynes.

Bates, A. W. (1982) Learning from audio-visual media: the Open University experience, *Institutional Research Review*, no. 1, pp. 33–58.

Bates, A. W. (1985) Using television in distance teaching, video pack presented at International World Conference of the Institute for Distance Education, Melbourne.

Bates, A. W. (1987) Teaching, media choice and cost effectiveness of alternative delivery systems, *Distance Education in Vocational Training*, no. 1, CEDEFOP, Berlin.

Bates, A. W. (1988a) Theory and practice in the use of technology in distance education, in D. Keegan (ed.) *Theoretical Principles of Distance Education*, Routledge, London and New York.

Bates, A. W. (1988b) Television, learning and distance education, *Journal of Educational Television*, Vol. 14, no. 3, pp. 213–226.

Bates, A. W. (1990) Audiocassettes in the British Open University, in A. W. Bates (ed.) *Media and Technology in European Distance Education*, Open University, Milton Keynes.

Bates, A. W. (1991) Third generation distance education: the challenge of new technology, *Research in Distance Education*, Vol. 3, no. 2, pp. 10–15.

Becker, H. S., Geer, B. and Hughes, E. C. (1968) *Making the Grade; the Academic Side of College Life*, Wiley, New York.

Beijderwellen, W. (1990) Interactive video in geology, in A. W. Bates (ed.) *Media and Technology in European Distance Education*, Open University, Milton Keynes.

Belsey, C. (1988) Marking by number, *AUT Woman*, no. 15, pp. 1–2.

Bentley, D. and Watts, D. M. (1989) *Learning and Teaching in School Science*, Open University Press, Milton Keynes.

Bergquist, W. (1993) *The Postmodern organization; Mastering the Art of Irreversible Change*, Jossey-Bass, San Francisco.

Beudoin, M. (1990) The instructor's changing role in distance education, *The American Journal of Distance Education*, Vol. 4, no. 2, pp. 21–29.

Black, P. J. and Ogborn, J. (1979) Laboratory work in undergraduate teaching, in D. McNally (ed.) *Learning Strategies in University Science*, University College Press, Cardiff.

Bloom, B. S. (ed.) (1956) *Taxonomy of Educational Objectives: Handbook One: the Cognitive Domain*, Wiley, New York.

Blowers, A. (1979) Carry on course teams, *Teaching at a Distance*, no. 16, pp. 54–57.

Bolton, J.P.R., Every, I. and Ross, S. M. (1990) The water videodisc: a problem-solving environment, *Computers and Education*, Vol. 15, nos. 1–3, pp. 165–172.

Bradley, C. (1984) Sex bias in the evaluation of students, *British Journal of Social Psychology*, Vol. 23, pp. 147–163.

Briggs, L. J. (1970) *Handbook of Procedures for the Design of Instruction*, American Institutes for Research, Pittsburgh.

Brown, A. L. (1988) Motivation to learn and understand: on taking charge of one's own learning, *Cognition and Instruction*, Vol. 5, no. 4, pp. 311–321.

Brown, D. and Clement, J. (1987) Overcoming misconceptions in mechanics: a comparison of two example based strategies. Paper presented at the annual meeting of the American Educational Research Association, Washington, DC.

Brown, J.S., Collins, A. and Duguid, P. (1989) Situated cognition and the culture of learning, *Educational Researcher*, Vol. 18, no. 1, pp. 32–42.

Brumby, M. (1984) Misconceptions about the concept of natural selection by medical biology students, *Science Education*, Vol. 68, no. 4, pp. 379–426.

Bruner J.S. (1960) *The Process of Education*, Harvard University Press, Cambridge, MA.

Bruner, J. S. (1961) The act of discovery, *Harvard Educational Review*, Vol. 31, no. 1, pp. 22–32.

Bruner, J.S. (1966) *Towards a Theory of Instruction*, Belknap Press, Cambridge, MA.

Bruner, J.S. (1985) Vygotsky: a historical and conceptual perspective, in J. V. Wertsz (ed.) *Culture, Communication and Cognition*, Cambridge University Press, Cambridge.

Bruner, J.S. (1986) *Actual Minds, Possible Worlds*, Harvard University Press, Cambridge, MA.

Bryce, T.G. and Robertson, I.J. (1985) What can they do? A review of practical assessment in science, *Studies in Science Education*, Vol. 12, pp. 1–24.

Burbules, N. and Linn, M. (1988) Response to contradiction: scientific reasoning during adolescence, *Journal of Educational Psychology*, Vol. 80, no. 1, pp. 61–75.

Butcher, P.G. and Greenberg, J.M. (1991) Educational computing at the Open University: the next decade, *Education and Computing*, Vol. 8, no. 1, pp. 201–215.

Butts, D. (1981) Keynote conference address on distance learning and broadcasting, in F. Percival and H. Ellington (eds.) *Aspects of Educational Technology XV: Distance Learning and Evaluation*, Kogan Page, London.

Buzan, T. (1974, revised 1989) *Use Your Head*, BBC Books, London.

Buzan, T. (1993) *The Mind Map Book*, BBC Books, London.

Calder, J. (1994) Course evaluation; improving academic quality and teaching effectiveness, in F. Lockwood (ed.) *Materials Production in Open and Distance Learning*, Paul Chapman Publishing, London.

Carey, S. (1986) Cognitive science and science education, *American Psychologist*, Vol. 41, no. 10, pp. 1123–1130.

Cassells, J. R. T. and Johnstone, A. H. (1983) The meaning of words and the teaching of chemistry, *Education in Chemistry*, Vol. 20, no. 1, pp. 10–11.

Chambers, E. A. (1989) Student workload and how to assess it, Open University Institute of Educational Technology internal report presented at Arts Faculty conference, Milton Keynes.

Chambers, E. A. (1992) Workload and the quality of student learning, *Studies in Higher Education*, Vol. 17, no. 2, pp. 141–152.

Chambers, E. A. (1994) Assessing learner workload, in F. Lockwood (ed.) *Materials Production in Open and Distance Learning*, Paul Chapman Publishing, London.

Champagne, A., Klopfer, L. and Anderson, J. (1980) Factors influencing the learning of classical mechanics, *American Journal of Physics*, Vol. 48, pp. 1074–1075.

Clark, R. (1983) Reconsidering research on learning from media, *Review of Educational Research*, Vol. 53, pp. 445–449.

Clark, R. E. (1990) When teaching kills learning: research on mathemathantics, in H. Mandle, E. DeCorte, S. N. Gennett and H. F. Friedrich (eds.) *Learning and Instruction: European Research in an International Context*, Vol. 2:2, *Analysis of Complex Skills and Complex Knowledge Domains*, Pergamon Press, New York.

Clement, J. (1982) Students' preconceptions in introductory mechanics, *American Journal of Physics*, Vol. 50, no. 1, pp. 66–71.

Clyde, A., Crowther, H., Patching, W., Putt, I. and Store, R. (1983) How students use distance teaching materials: an institutional study, *Distance Education*, Vol. 4, no. 1, pp. 4–26.

Coffey, J. (1977) in T. C. Davies (ed.) *Open Learning Systems for Mature Students*, Working Paper 14: Council for Educational Technology, London.

Collins, A. (1993) A design science of education, in T. O'Shea and E. Scanlon (eds.) *op. cit.*

Convery, A. (1990) Using television in the classroom: a teacher's experience, *Journal of Educational Television*, Vol. 16, no. 3, pp. 151–162.

Cornish, J. C. L., Jenkins, P. J. and Atkinson, R. J. (1981) External studies in undergraduate physics: self-paced learning with laboratory kits, *Distance Education*, Vol. 2, no. 2, pp. 137–145.

Costello, N. (1979) The curse of the course team: a comment, *Teaching at a Distance*, no. 16, pp. 53–54.

Cowan, J. (1992) in M-SCAPE, Edition no. 2 (a newsletter of the UKOU Mathematics Faculty for tutorial and counselling staff), Open University internal publication, Milton Keynes.

Crooks, B., Rowntree, D. and Waller, R. (1979) Writing your lesson, in D. Rowntree and B. Connor (eds.) *How to Develop Self Instructional Teaching*, OUCICS, Milton Keynes.

Crooks, B. and Kirkwood, A. D. (1988) Videocassettes by design in OU courses, *Open Learning*, Vol. 3, no. 3, pp. 13–17.

Crooks, B. and Kirkwood, A. D. (1990) Videocassettes by design in Open University courses, in A. W. Bates (ed.) *Media and Technology in European Distance Education*, Open University, Milton Keynes.

Crowther, G. (1971) Inaugural address (1969) of the first Chancellor of the OU, reproduced in the *Undergraduate Prospectus 1971*, and reprinted in Open University 25th Anniversary Service of Thanksgiving (1994), p. 3, Open University Press, Milton Keynes.

CTISS (1992) *Computers in University Teaching: Core Tools for Core Activities*, CTISS Publications, Oxford.

Curran, C. (1990) Factors affecting the costs of media in distance education, in A. W. Bates (ed.) *Media and Technology in European Distance Education*, Open University, Milton Keynes.

Curriculum Development Centre (1988) *Science for Everybody? Towards a National Science Statement*, Curriculum Development Centre, Canberra.

Daniel, J. (1988) The worlds of open learning, in N. Paine, (ed.) *op. cit.*

Daniel, J. S. and Marquis, C. (1977) Interaction and independence: getting the mixture right, *Teaching at a Distance*, no. 14, pp. 29–44.

Darby, J. (1989) Personal communication of quote from Ellis at the CAL 89 conference, April.

Darby, J. (1992) Editorial, *The CTISS File*, no. 14, p. 2.

Darby, J. (1993) Editorial, *The CTISS File*, no. 15, April, p. 1.

Darby, J. (1995) Material case for mixed-up culture, *The Times Higher Educational Supplement, Multimedia Features*, 10 March, p. vi.

Davies, G. M. and Nisbet, J. (1981) Cognitive psychology and curriculum development, *Studies in Science Education*, Vol. 8, pp. 127–134.

Davies, K. (1994) Teaching in the postmodern classroom, presentation given at the Eleventh International Conference on Technology and Education, London, March.

Department of Education and Science, The National Curriculum Council (1985) *Science in Schools at age 13 and 15* HMSO, London.

Department of Education and Science (1988) *Task Group on Assessment and Testing: a Report*, DES, London.

Department of Education and Science, The National Curriculum Council (1989) *Science in the National Curriculum*, HMSO, London.

Derycke, A. C. and Kaye, A.R. (1993) Participative modelling and design of collaborative distance learning tools, in G. Davies and B. Samways (eds.) *Teleteaching: Proceedings of the IFIP Third Teleteaching Conference*, IFIP, North-Holland, Amsterdam.

Dewey, J. (1916) *Democracy in Education*, Macmillan, New York.

Di Sessa, A. (1988) Knowledge in pieces, in G. Forman and P. Pufall (eds.) *Constructivism in the Computer Age*, pp. 49–70, Lawrence Erlbaum Associates, Hillsdale, NJ.

Dodds, T. (1993) Procedures for materials and text development in distance education, paper prepared for the First Korea Air and Correspondence University Conference, Seoul, Korea, International Extension College, Cambridge.

Drake, M. (1979) The curse of the course team, *Teaching at a Distance*, no. 16, pp. 50–53.

Draper, S., McAtteer, E., Tolmie, A. and Anderson, A. (1994) Bringing a constructivist approach to bear on evaluation, in H. Foot, C. Howe, A. Anderson, A. Tolmie and D. Warden (eds.) *Group and Interactive Learning*, Computational Mechanics Publications, Southampton.

Driver, R. and Easley, J. (1978) Pupils and paradigms: a review of literature related to concept development in adolescent science students, *Studies in Science Education*, Vol. 5, pp. 61–84.

Driver, R. (1984) Cognitive psychology and pupils' frameworks in mechanics, in *Proceedings of the International Conference on Physics Education: The Many Faces of Teaching and Learning Mechanics*, Utrecht.

Driver, R., Guesné, E. and Tiberghien, A. (eds.) (1985) *Children's Ideas in Science*, Open University Press, Milton Keynes.

Duffy, T. (1985) Readability formulas: what's the use?, in T. Duffy and R. M. Waller (eds.) *Designing Usable Texts*, Academic Press, Orlando, FL.

Dunbar, R.I.M. (1982) Open and closed science; and Open University tutoring, *Teaching at a Distance*, Vol. 21, pp. 68–70.

Durbridge, N. (1980) S101 audio cassettes, Institute of Educational Technology paper on broadcasting, no. 139, Open University internal document, Milton Keynes.

Durbridge, N. (1983) Design implication of audio and video cassettes, Institute of Educational Technology paper on broadcasting, no. 222, Open University internal document, Milton Keynes.

Durbridge, N. (1984) Developing the use of videocassettes at the Open University, in O. Zuber-Skerritt (ed.) *Video in Higher Education*, Kogan Page, London.

Edwards, D. and Mercer, N. (1987) *Common Knowledge*, Routledge, London.

Edwards, D. J. (1979), Some factors affecting the reliability of tutor grades at the Open University, *Assessment in Higher Education*, Vol. 5, no. 1, pp. 16–44.

Edwards, D. J. (1982) Project marking: some problems and issues, *Teaching at a Distance*, Vol. 21, pp. 28–34.

Eicher, J-C., Hawkridge, D., McAnany, E., Mariet, F. and Orivel, F. (1982) *The Economics of New Educational Media*, Vol. 3, UNESCO Press, Paris.

Eisenstadt, M. (1994) *The Virtual Summer School*, World Wide Web.

Eisenstadt, M. (1995) Overt strategies for global learning, *The Times Higher Educational Supplement, Multimedia Features*, 7 April, pp. vi–vii.

Eisenstadt, M., Brayshaw, M., Hasemer, T. and Issroff, K. *Teaching, Learning and Collaborating at an Open University Virtual Summer School* (in press).

Eisner, E. W. (1985) *The Art of Educational Evaluation*, Falmer Press, London and Philadelphia.

Ellerman, H. H., Huisman, W. H. T., Scellekens, A. M. H. C., Zwaneveld, G. and Berns, R. M. (1992) An experimental network-mediated study support system using computer-mediated communication as a focus for tutor-student communication, *Journal of Computer Assisted Learning*, Vol. 8, pp. 186–192.

Engel-Clough, E. and Driver, R. (1986) A study of consistency in the use of students' conceptual frameworks across two different task frameworks, *Science Education*, Vol. 70, no. 4, pp. 473–496.

Entwistle, A. and Entwistle, N. (1992) Experiences of understanding in revising for degree examinations, *Learning and Instruction*, Vol. 2, no. 1, pp. 1–22.

Entwistle, N. and Entwistle, A. (1991) Contrasting forms of understanding for degree examinations: the student experience and its implications, *Higher Education*, Vol. 22, no. 3, pp. 205–228.

Eraut, M. (1972) Strategies for the evaluation of curriculum materials, in K. Austwick and N. Harris (eds.) *Aspects of Educational Technology VI*, Pitman, London.

ETS Educational Testing Service (1991) *How can We Judge the Fairness of Tests?* Public Accountability Report, ETC, Princeton, NJ.

Evans, T. and Nation, D. (1989) *Critical Reflections on Distance Education*, Falmer Press, Lewes.

Evans, T. (1993) *Understanding Learning in Open and Distance Education*, Kogan Page, London.

Every, I. and Scanlon, E. (1983) Discovering physics with microcomputers, *Computers in Education*, Vol. 8, no. 1, pp. 183–188.

Fairbrother, R. (1988a) Problems in the assessment of scientific skills, in J. Wellington (ed.) *Skills and Processes in Science Education*, Routledge, London.

Fairbrother, R. (1988b) Principles of practical assessment, in B. Woolnough (ed.) *Practical Science*, Open University Press, Milton Keynes.

Feldsine, J. E. Jr. (1987) The construction of concept maps facilitates the learning of general college chemistry: a case study, unpublished doctoral dissertation, Department of Education, Cornell University, Ithaca, New York.

Fensham, P. J. (1985) Science for all, *Journal of Curriculum Studies*, Vol. 17, no. 4, pp. 415–435.

Fensham, P. J. (1988a) Familiar but different: some dilemmas and new directions in science education, in P. Fensham (ed.) *op. cit.*

Fensham, P. (ed.) (1988b) *Development and Dilemmas in Science Education*, Falmer Press, Lewes.

Fensham, P. (1992) Science and technology, in P. Jackson (ed.) *Handbook of Research on Curriculum*, pp. 789–829, Macmillan, New York.

Fensham, P. J. (1993) Reflections on science for all, in E. Whitelegg, J. Thomas and S. Tresman (eds.) *op. cit.*

Field, J. (1982) Student learning from media: student diversity versus the centralised institution, *Institutional Research Review*, Vol. 1, pp. 101–135.

Fielden, J. and Pearson, P. K. (1978) *The Cost of Learning with Computers*, Council for Educational Technology, London.

Fowler, P. (1995) Challenge to courseware culture club, *The Times Higher Educational Supplement, Multimedia Features*, 10 March, p. x.

Freeman, R. (1979) 'Botanic man': education or entertainment? *Teaching at a Distance*, no. 16, pp. 19–23.

Gagné, R. M. (1965) *The Conditions of Learning*, Holt, Rinehart & Winston, New York.

Gagné, R. and Briggs, L. (1974) *Principles of Instructional Design*, Holt, Rinehart & Wilson, New York.

Gagné, R. M. (1985) *The Conditions of Learning* (4th edn), Holt, Rinehart & Winston, New York.

Gallagher, M. A. (1977) Broadcasting and the Open University student, Institute of Educational Technology papers on broadcasting, no. 80, Open University internal report, Milton Keynes.

Gallagher, M. A. (1978) Good television and good teaching – some tensions in educational practice, *Educational Broadcasting International*, Vol. 11, no. 3; also IET paper on broadcasting, no. 91, Open University internal publication, Milton Keynes.

Gardiner, J. M. (1989) A generation effect in memory without awareness, *British Journal of Psychology*, Vol. 80, pp. 163–168.

Garg, S., Vijayshre and Panda, S. (1992) A preliminary study of student workload for IGNOU elective physics courses, *Indian Journal of Open Learning*, Vol. 1, no. 2, pp. 19–25.

Garrison, D. R. (1986) Multifunction microcomputer enhanced audio teleconferencing: moving into the third generation of distance education, *International Journal of Innovative Higher Education*, Vol. 3, no. 1, pp. 26–29.

Gibbs, G. and Durbridge, N. (1976) Characteristics of Open University tutors (Part 2): tutors in action, *Teaching at a Distance*, no. 7, pp. 7–22.

Gibbs, G. (1981) *Learning to Study*, National Extension College, Cambridge.

Gibbs, G. (1984) Learning to learn – the student-centered approach, in E. S Henderson and M. B. Nathenson (eds.), *op. cit.*

Giddings, G., Hofstein, A. and Lunetta, V. (1988) Assessment and evaluation in practical science, in B. Woolnough (ed.) *Practical Science*, Open University Press, Milton Keynes.

Gipps, C. and Murphy, P. (1994) *A Fair Test?*, Open University Press, Milton Keynes.

Glaser, R. (1984) Education and thinking: the role of knowledge, *American Psychologist*, Vol. 39, no. 2, pp. 93–104.

Glover, D. M., Graham, G. R. and MacDonald, R. M. (1989) The CCAT videodisc – a new resource for physics education, *Physics Education*, Vol. 24, pp. 304–308.

Goodhart, C. B. (1992) Sex and class in examinations, *The Cambridge Review*, Vol. 113, pp. 43–44.

Gott, R. and Murphy, P. (1987) Assessing investigations at ages 13 and 15, *Science Report for Teachers, 9*, ASE, Hatfield.

Granger, D. (1990) Open learning and individualised distance learning at Empire State College, *Open Learning*, Vol. 5, no. 1, pp. 24–30.

Greenberg, J. (1995) On the open highway in search of serious money, *The Times Higher Educational Supplement, Multimedia Feature*, 7 April, p. vii.

Greenfield, D. (1984) Home kits, in A. W. Bates (ed.) *The Role of Technology in Distance Education*, Croom Helm, London.

Gunning, R. (1952) *The Technique of Effective Writing*, McGraw Hill, New York.

Gunter, B. (1985) *Dimensions of Television Violence*, Gower, London.

Gunter, B. and McLaughlin, C. (1992) *Television: The Public's View*, Independent Television Commission Research Monograph, John Libbey, London.

Haight, G. and Jones, L. (1987) Kinetics and mechanisms of the iodine azide reaction: a videotaped experiment, *Journal of Chemical Education*, Vol. 64, no. 3, pp. 271–273.

Harlen, W. (1992) Research and the development of science in the primary school, *International Journal of Science Education*, Vol. 14, no. 5, pp. 491–503.

Harlen, W. (1993) Education for equal opportunities in a scientifically literate society, in E. Whitelegg, J. Thomas and S. Tresman (eds.) *op. cit.*

Harris, D. and Bell, C. (1986) *Evaluation and Assessing for Learning*, Kogan Page, London.

Harris, D. (1987) *Openness and Closure in Distance Education*, Falmer Press, Lewes.

Harry, K., John, M. and Keegan, D. (eds.) (1993) *Distance Education: New Perspectives*, Routledge, London and New York.

Hartley, J. and Burnhill, P. (1977a) Fifty guide-lines for improving instructional text, *Programmed Learning and Educational Technology*, Vol. 14, pp. 65–73.

Hartley, J. and Burnhill, P. (1977b) Understanding instructional text: typography, layout, design, in M. J. A. Howe (ed.) *Adult Learning; Psychological Research and Applications*, Wiley, London.

Hartley, J. R., Mallen, C. and Byard, M. (1991) Qualitative modelling in cognitive change in science instruction, *Proceedings of the First International Conference on the Learning Sciences*, pp. 223–236.

Hartog, P. and Rhodes, E. C. (1936) *The Marks of Examiners*, Macmillan, London.

Hawkridge, D., Vincent, T. and Hales, G. (1985) *New Information Technology in the Education of Disabled Children and Adults*, Croom Helm, London.

Hawkridge, D. G. (1990) Creative gales and computers in third world schools, in O. Boyd-Barrett and E. Scanlon (eds.) *Computers and Learning*, Addison Wesley, Wokingham.

Hawkridge, D. G. (1991) *Evaluating the Cost Effectiveness of Advanced IT and Learning*, paper presented at the workshop on advanced IT and learning, Open University, Milton Keynes, 5–6 November.

Hawkridge, D. (1994) Which team for open and distance learning materials production?, in F. Lockwood (ed.) *op. cit.*

Helm, H. and Novak, J. (1983) *Proceedings of the International Seminar: Misconceptions in Science and Mathematics*, Cornell University, Ithaca.

Henderson, E., Hodgson, B. and Nathenson, M. (1977) Developmental testing: the proof of the pudding, *Teaching at a Distance*, no. 10, pp. 77–82.

Henderson, E., Hodgson, B., Kirkwood, A., Lefrere, P., Mace, E., Mayer, B. and Nathenson, M. (1980) Developmental testing for credit, Open University internal report, Milton Keynes.

Henderson, E. S. and Nathenson, M. B. (eds.) (1984) *Independent Learning in Higher Education*, Educational Technology Publications, Englewood Cliffs, NJ.

Henry, N. W. (1975) Objectives of laboratory work, in P. Gardner (ed.) *The Structure of Science Education*, Longman, Australia.

Hewson, M. and Hewson, P. (1983) The effect of instruction using students' prior knowledge and conceptual change categories on science learning, *Journal of Research on Science Teaching*, Vol. 20, no. 2, pp. 731–743.

Hiltz, R. (1984) *Online Communities: a Case Study of the Office of the Future*, Ablex Publishing Corporation, NJ.

HMSO (1960) *Science in Secondary Schools*, Ministry of Education Pamphlet No. 38, HMSO, London.

HMSO (1966) *A University of the Air*, White Paper presented by the Secretary of State for Education and Science.

Hodgson, D. (1984) Learning from lectures, in F. Marton, D. J. Hounsell and N. J. Entwistle (eds.) *The Experience of Learning*, Scottish Academic Press, Edinburgh.

Hudson, L. (1966) *Contrary Imaginations*, Methuen, London.

Issroff, K. (1994a) Gender and cognitive and affective aspects of cooperative learning, in H. Foot, C. Howe, A. Anderson, A. Tolmie and D. Warden (eds.) *Group and Interactive Learning*, Computational Mechanics Publications, Southampton.

Issroff, K. (1994b) Questionnaire reports of students on the Virtual Summer School, Computer Assisted Learning, Technical Report, Open University internal document, Milton Keynes.

Jacobs, G. (1992) Remote interactive multimedia: a necessary condition, *The CTISS File*, no. 14, pp. 12–15.

Jenkins, D. A. (1987) The use of computers in school biology, *School Science Review*, Vol. 68, no. 645, pp. 687–693.

Jenkins, E. W. (1987) Philosophical flaws, *The Times Educational Supplement*, 2 January.

Jenkins, S. (1995) Whingeing has become a habit of mind for British science, *The Spectator*, 25 March.

Johnson, S. and Murphy, P. (1986) *Girls and Physics*, DES, London.

Johnstone, A. H. (1991) Why is science difficult to learn? Things are seldom what they seem, *Journal of Computer Assisted Learning*, Vol. 7, no. 2, pp. 75–83.

Jonassen, D., Mayes, T. and McAleese, R. (1993) A manifesto for a constructivist approach to uses of technology in higher education, in T. Duffy, J. Lowyck and D. Jonassen (eds.) *Designing Environments for Constructive Learning*, Springer Verlag, New York.

Jonassen, D. (1991) Hypertext as instructional design, *Educational Technology Research and Development*, Vol. 38, no. 1, pp. 83–92.

Jones, A. and O'Shea, T. (1982) An evaluation of tutorial CAL at the Open University, *British Journal of Educational Technology*, Vol. 13, no. 3, pp. 207–217.

Jones, L.L., (1988) Enhancing instruction in the practice of chemistry with the computer-assisted interactive videodisc, *Trends in Analytical Chemistry*, Vol. 7, no. 8, pp. 273–276.

Kandaswamy, S. (1980) Evaluation of instructional materials: a synthesis of models and methods, *Educational Technology*, Vol. 20, no. 10, pp. 19–26.

Kaye, A. (1973) Design and evaluation of science courses at the Open University, *Instructional Science*, Vol. 2, pp. 119–185.

Kaye, A. and Pentz, M. J. (1974) Integrating multi-media systems for science education which achieve a wide territorial coverage, in *New Trends in the Utilisation of Educational Technology for Science Education*, UNESCO Press, Paris.

Kaye, A. (1981) Characteristics of distance-learning systems, in A. Kaye and G. Rumble (eds.) *op. cit.*

Kaye, A. and Rumble, G. (eds.) (1981) *Distance Teaching for Higher and Adult Education*, Croom Helm, London.

Kaye, A., Mason, R. and Harasim, L. (1989) Computer conferencing in the academic environment, CITE Report no. 91, Open University internal publication, Milton Keynes.

Kaye, A. (1990) Computer conference and mass distance education, CITE Report no. 98, Open University internal publication, Milton Keynes.

Kaye, A. (1991) Learning together apart, in A. Kaye (ed.) *Collaborative Learning through Computer Conferencing*, Springer Verlag, Heidelberg.

Keegan, D. (1983) On defining distance education, in D. Sewart, D. Keegan and B. Holmberg (eds.) *op. cit.* (reprint of 1980 article originally published in *Distance Education*, Vol. 1, no. 1, pp. 13–36).

Keegan, D. (1990) *Foundations of Distance Education* (2nd edn.), Routledge, London.

Kember, D. and Murphy, D. (1990) A synthesis of open, distance and student-centred learning, *Open Learning*, Vol. 5, no. 2, pp. 3–8.

Kemmis, S., Atkins, R. and Wright, E. (1977) *How Do Students Learn: the UNCAL Evaluation*, occasional publication no. 5, Centre for Applied Research in Education, University of East Anglia, Norwich.

Kirk, P. (1976) The loneliness of the long-distance tutor, *Teaching at a Distance*, no. 7, pp. 3–6.

Kirkup, G. (1981) Evaluating and improving learning materials: a case study, in F. Percival and H. Ellington (eds.) *Aspects of Educational Technology, XV*, pp. 164–171, Kogan Page, London.

Kirkup, G. (1984) Teaching study skills in context, in E. S. Henderson and M.B. Nathenson (eds.), *op. cit.*

Kirkwood, A. D. (1990) Into the video age: Open University television in the 1990s, *Journal of Educational Television*, Vol. 16, no. 2, pp. 77–86.

Kirkwood, A. D. (1991) Students and science television, Programme on Learner Use of Media, paper no. 28, Open University internal document, Milton Keynes.

Kirkwood, A. (1992) Assessment at the Open University, Student Research Centre report no. 59, Open University internal document, Milton Keynes.

Kirschner, P. A. and Meester, M. A. M. (1988) The laboratory in higher science education, *Higher Education*, Vol. 17, pp. 81–98.

Klare, G. R. (1981) Readability indices: do they inform or misinform? *Information Design Journal*, Vol. 2, no. 3/4, pp. 251–255.

Krol, E. (1994) *The Whole Internet: User's Guide and Catalog* (2nd edn), O'Reilly and Associates, Sebastopol, CA.

Laurillard, D. (1979) The processes of learning, *Higher Education*, Vol. 8, pp. 359–409.

Laurillard, D. (1987) Problems and possibilities of interactive video, in A. Jones, E. Scanlon and T. O'Shea (eds.) *The Computer Revolution in Education*, Harvester Press, Brighton.

Laurillard, D., Swift, B. and Darby, J. (1992) Probing the not invented here syndrome, *The CTISS File*, no. 14, p. 54.

Laurillard, D. (1993) *Rethinking University Teaching*, Routledge, London.

Lauzon, A. C. and Moore, G. A. B. (1989) A fourth-generation distance education system, *American Journal of Distance Education*, Vol. 3, no. 1, pp. 38–49.

Lawless, C. (1989) Report on the annual survey of new courses, IET internal document, Open University, Milton Keynes.

Lawless, C. (1994) Course design: order and presentation, in F. Lockwood (ed.) *op. cit.*

Lentell, H. (1994) Why is it so hard to hear the tutor in distance education? *Open Learning*, Vol. 9, no. 3, pp. 49–52.

Lewis, B. (1972) Course production at the Open University IV: the problem of assessment, *British Journal of Educational Technology*, Vol. 2, no. 3, p. 109.

Lewis, B. and Faziollahi, B. (1989) A fresh look at an old topic: testing, *Connexions*, Vol. 2, no. 2.

Lewis, E. (1990) The development of understanding in elementary thermodynamics: a study of conceptual change and the factors that affect that change, unpublished doctoral dissertation, University of California, Berkeley.

Lewis, R. (1976) *How to Write Essays*, National Extension College, Cambridge.

Lewis, R. (1986) What is open learning? *Open Learning*, Vol. 1, no. 2, pp. 5–10.

Lewis, R. and Spencer, D. (1986) *What is Open Learning?* Open Learning Guides series, Vol. 4, Council for Educational Technology, London.

Lewis, R. (1988) Open learning – the future, in N. Paine (ed.) *op. cit.*

Linn, M. and Pulos, S. (1988) Male–female differences in predicting displaced volume: strategy, usage, aptitude relationships and experience influences, *Journal of Educational Psychology*, Vol. 75, pp. 86–96.

Linn, M. and Songer, N. (1988). Cognitive and conceptual change across adolescence, *American Journal of Education*, Vol. 99, no. 4, pp. 379–417.

Linn, M., Songer, N., Lewis, E. and Stern, J. (1991) Using technology to teach thermodynamics: achieving integrated understanding, in D. Ferguson (ed.) *Advanced Educational Technologies for Mathematics and Science*, Springer Verlag, Berlin.

Linn, M. and Songer, N. (1993) reprinted in D. Edwards, E. Scanlon and R. West (eds.) *Teaching, Learning and Assessment in Science Education*, Paul Chapman Publishing, London.

Lockwood, F. (ed.) (1992) *Activities in Self-Instructional Texts*, Kogan Page, London.

Lockwood, F. (ed.) (1994) *Materials Production in Open and Distance Learning*, Paul Chapman Publishing, London.

Lucas, A. (1987) Public knowledge of biology, *Journal of Biological Education*, Vol. 20, pp. 41–45.

Lunneborg, P. (1994) *OU Women: Undoing Educational Obstacles*, Cassell, London.

Macdonald-Ross, M. (1973) Behavioural objectives – a critical review, *Instructional Science*, Vol. 2, pp. 1–52.

Macdonald-Ross, M. and Waller, R. H. W. (1975) Criticisms, alternatives and tests: a framework for improving typography, *Programmed Learning and Educational Technology*, Vol. 12, no. 2, pp 75–83.

Macdonald-Ross, M. (1977) Graphics in texts, *Research in Graphic Communication*, Institute of Educational Technology Monograph no. 7, Open University, Milton Keynes.

Mackenzie, N., Postgate, R. and Scupham, J. (1975) *Open Learning: Systems and Problems in Post Secondary Education*, UNESCO Press, Paris.

Mager, R. F. (1961) On the sequencing of instructional content, *Psychological Reports*, no. 9, pp. 405–413.

Mager, R. F. (1962) *Preparing Instructional Objectives*, Fearon, Belmont, CA.

Malone, T. W. (1981) Towards a theory of intrinsically motivating instruction, *Cognitive Science*, Vol. 4, pp. 33–69.

Marchioni, G. and Schneiderman, B. (1988) Finding facts versus browsing knowledge in hypertext systems, *Computer*, Vol. 21, no 1, pp. 70–80; reprinted in O. Boyd-Barrett and E. Scanlon (eds.) (1990) *Computers and Learning*, Addison Wesley, Wokingham.

Marland, P. W. and Store, R. E. (1982) Some instructional strategies for improved learning from distance teaching materials, *Distance Education*, Vol. 3, no. 1, pp. 72–106.

Marland, P. W., Patching, W., Putt, I. and Putt, R. (1990) Distance learners' interactions with text while studying, *Distance Education*, Vol. 11, no. 1, pp. 71–91.

Marton, F. and Säljö, R. (1976a) On qualitative differences in learning I: outcome and processes, *British Journal of Educational Psychology*, Vol. 46, pp. 4–11.

Marton, F. and Säljö, R. (1976b) On qualitative differences in learning II: outcome as a function of the learners' conception of the task, *British Journal of Educational Psychology*, Vol. 46, pp. 115–127.

Mason, J. (1976) Life inside the course team, *Teaching at a Distance*, no. 5, pp. 27–33.

Mason, J. (1985) Changing attitudes to women in science and technology in Great Britain; the Open University experience, in E. Vamos Uvesco (ed.) *Women in Science: Options and Access*, UNESCO, Budapest.

Mason, R. (1989a) An evaluation of CoSy on an Open University course, in R. Mason and A. Kaye (eds.) *Mindweave*, Pergamon, Oxford.

Mason, R. (1989b) *The Use of Computer Networks for Education and Training*, Training Agency, Sheffield.

Mason, R. and Kaye, A. (1989) *Mindweave: Communication, Computers and Distance Education*, Pergamon, Oxford.

Mason, R. (1995) The MINT project in the context of future directions for computer conferencing in the Open University, CITE report no. 207, Open University internal report, Milton Keynes.

Maugh, T. (1985) A new trend: training on the 'Tube', *Science*, Vol. 227, no. 4694, pp. 1569–1570.

Mayer, R. E. (1979) Can advance organisers influence meaningful learning, *Review of Educational Research*, Vol. 4, no. 2, pp. 371–383.

Mayer, R. E. and Gallini, J. K. (1990) When is an illustration worth ten thousand words? *Journal of Education Psychology*, Vol. 82, no. 4, pp. 715–726.

McConnell, D. (1992) Computer mediated communication for management learning, in A. R. Kaye (ed.) *Collaborative Learning through Computer Conferencing*, NATO ASI Series F: Computer and System Sciences, Vol. 90, Springer Verlag, Berlin.

McConnell, D. (1994) *Implementing Computer Supported Cooperative Learning*, Kogan Page, London.

McCormick, R. (1976) Evaluation of Open University course materials, *Instructional Science*, Vol. 5, pp. 189–217.

McCulloch (1994) in A. Assiter and K. Swann (eds.) *op. cit.*

McDermott, L. (1984) Research on conceptual understanding in mechanics, *Physics Today*, Vol. 37, July, pp. 24–32.

McGrath, J.E. (1990) Time matters in groups, in J. Galegher, R. Kraut and C. Egido (eds.) *Intellectual Teamwork: Social and Technological Foundations of Cooperative Work*, Lawrence Erlbaum, Hillsdale, NJ.

McIntosh, N. (1974) The OU student, in J. Tunstall (ed.) *The Open University Opens*, Routledge & Kegan Paul, London.

Megarry, J. (1989) Hypertext and compact discs: the challenge of multimedia learning, *Aspects of Educational and Training Technology XXII: Promoting Learning*, Kogan Page, London.

Melton, R. (1976) A guide to item analysis, Open University internal document, Milton Keynes.

Melton, R. (1978) Resolution of conflicting claims concerning the effect of behavioral objectives on student learning, *Review of Educational Research*, Vol. 48, pp. 291–302.

Melton, R. (1982) *Instructional Models for Course Design and Development*, Educational Technology Publications, NJ.

Melton, R. and Scanlon, E. (1982) Learning from evaluation for S101, *Institutional Research Review*, no. 2, pp. 61–72.

Melton, R. (1984) Alternative forms of preliminary organiser, in E. Henderson and M. B. Nathenson (eds.) *Independent Learning in Higher Education*, Educational Technology Publications, NJ.

Melton, R. F. (1990) Transforming text for distance learning, *British Journal of Educational Technology*, Vol. 21, no. 3, pp. 183–195.

Merril, M. D. (1987) The new component design theory: instructional design for courseware authoring, *Instructional Science*, Vol. 16, no. 1, pp. 19–34.

Merril, M. D. (1991) Constructivism and instructional design, *Educational Technology*, Vol. 21, no. 5, pp. 45–53.

Millar, R. and Driver, R. (1987) Beyond process, *Studies in Education*, Vol. 14, pp. 33–62.

Millar, R. (1989) in J. Wellington (ed.) *op. cit.*

Millar, R. (1991) Why is science hard to learn? *Journal of Computer Assisted Learning*, Vol. 7, no. 2, pp. 66–74.

Miller, C. and Parlett, M. (1976) Up to the mark: a research report on assessment, Occasional Paper no.13, Centre for Research in the Educational Sciences, University of Edinburgh, Edinburgh.

Miller, G. (1976) Continuous assessment, *Medical Education*, Vol. 10, pp. 81–86.

Minstrell, J. (1982) Explaining the 'at rest' condition of an object, *The Physics Teacher*, Vol. 20, pp. 10–14.

Monk, J. and O'Shea, T. (1981) Planning and role differentiation in course production, *Teaching at a Distance*, no. 19, pp. 62–66.

Morgan, A. (1993) *Improving Your Students' Learning*, Kogan Page, London.

Moses, D. and Croll, P. (1991) *Educational Television in Classrooms and Schools*, Bristol BRU/ERA/ Bristol Polytechnic, Bristol.

Murphy, P. (1988) TGAT; a conflict of purpose, *Curriculum*, Vol. 9, no. 3, pp. 152–158.

Murphy, P. (1991) Assessment, in K. Mercer, (ed.) *Genderwatch*, Cambridge University Press, Cambridge.

Murphy, P. and Scanlon, E. (1993) Perceptions of process and content in the science curriculum, in G. Bourne (ed.) *Thinking through Primary Practice*, Routledge, London.

Murphy, P. and Scanlon, E. (1995) Developing the teaching profession through distance learning: a multiple media approach, *Proceedings of the International Conference on Distance Learning*, Birmingham, June.

Nathenson, M. B. and Henderson, E. (1980) *Using Student Feedback to Improve Learning Materials*, Croom Helm, London.

Nathenson, M., Brown, S., Kirkup, G., Lewsey, M. and Spratley, I. (1981) Learning from evaluation at the Open University, *British Journal of Educational Technology*, Vol. 8, pp. 97–102.

National Science Foundation (1983) *Educating Americans for the 21st Century*, National Science Foundation, Washington, DC.

Nelson, T. (1967) Getting it out of our system, in G. Schecter (ed.) *Information Retrieval: A Critical View*, pp. 191–210, Thompson Book Company.

New Scientist (1985) What do people think of science? 21 Feb, no. 1444, pp. 12–16.

Nipper, S. (1989) Third generation distance learning and computer conferencing, in R. Mason and A. Kaye (eds.) *Mindweave*, Pergamon, Oxford.

Northedge, A. (1987) Address to a workshop of Open University regional academic staff, private communication, unpublished.

Northedge, A. (1990) *The Good Study Guide*, Open University, Milton Keynes.

Novak, J.D. (1990) Concept maps and vee diagrams: two meta-cognitive tools to facilitate meaningful learning, *Instructional Science*, Vol. 19, pp. 29–52.

Nunan, T. (1992) *Student Feedback for the Evaluation of Distance Teaching and Learning*, Distance Education Centre, University of South Australia, Underdale.

Nunan, T. and Calvert, J. (1992) *Quality and Standards in Distance Education*, report to the National Distance Education Conference, University of South Australia and Deakin University.

Nuttall, D. (1987) The validity of assessment, *European Journal of Psychology of Education*, Vol. 11, no. 2, pp. 109–118.

O'Shea, T. and Self, J. (1983) *Learning and Teaching with Computers*, Prentice Hall, Englewood Cliffs, NJ.

O'Shea, T., O'Malley, C. and Scanlon, E. (1990) Magnets, Martians and mathematical microworlds: learning with and learning by OOPS, *Journal of Artificial Intelligence in Education*, Vol. 1, no. 3, pp. 11–26.

O'Shea, T. and Scanlon, E. (eds.) (1993) *New Directions in Educational Technology*, Springer Verlag, Berlin.

O'Shea, T. (1995) Towards the virtual university, keynote address given at the CAL 95 conference, Cambridge, April.

Ogborn, J. and Wong, D. (1984) A microcomputer dynamic modelling system, *Physics Education*, Vol. 19, no. 3, pp. 138–142.

Ogborn, J. (1987) Computational modelling in science, in R. Lewis and E. D. Tagg (eds.) *Trends in Computer Assisted Learning*, Blackwell, Oxford.

Okebukola and Jegede (1988) Cognitive preference and learning mode as determinants of meaningful learning through concept mapping, *Science Education*, Vol. 72, no. 4, pp. 489–500.

Olson, D. and Bruner, J. (1974) Learning through experience and learning through media, in D. Olson (ed.) *Media and Symbols: the Forms of Expression* (73rd NSSE Yearbook), University of Chicago, Chicago.

Open University (1978) Unpublished report on redesigned exams and assessment system, OU, Milton Keynes.

Open University (1985) *Primary Science: Why and How* (course code EHP 531), OU, Milton Keynes.

Open University (1988) *Open Teaching*, a guide for tutorial and counselling staff, OU, Milton Keynes.

Open University (1991) Science Faculty Profile Review Group Interim Report, unpublished internal document, OU, Milton Keynes.

Open University (1992) Unpublished internal document, private communication, OU, Milton Keynes.

Open University (1995) Supplementary course material S271 PB, OU, Milton Keynes.

Owen, S. (1994) Oxbridge – the stuff study myths are made of, *Sesame* (Open University student newspaper), issue 163, p. 11.

Page, E.B. (1958) Teachers' comments and student performance, *Journal of Educational Psychology*, Vol. 49, pp. 173–181.

Pagney, B. (1980) Paper given at EHSC Conference (in French). English translation reprinted in D. Sewart, D. Keegan and B. Holmberg (eds.) *op. cit.*

Paine, N. (ed.) (1988) *Open Learning in Transition: An Agenda for Action*, National Extension College, Cambridge.

Papert, S. (1980) *Mindstorms: Children, Computers and Powerful Ideas*, Basic Books, New York.

Parlett, M. and Hamilton, D. (1972) Evaluation as illumination: a new approach to the study of innovatory programs, in D. Hamilton (ed.) *Beyond the Numbers Game: a Reader in Educational Evaluation*, Macmillan, London.

Parlett, M. and Woodley, A. (1983) Student drop out, *Teaching at a Distance*, Vol. 24, pp. 2–23.

Partington, P. (1995) A never-ending learning story, *The Times Higher Educational Supplement*, 20 January, p. 12.

Perraton, H. (1982a) *The Cost of Distance Education*, International Extension College, Cambridge.

Perraton, H. (ed.) (1982b) *Alternative Routes to Formal Education*, World Bank/Johns Hopkins University Press, Baltimore.

Perraton, H. (1983) A theory for distance education, in D. Sewart, D. Keegan and B. Holmberg (eds.) *op. cit.* (reprint of 1981 article originally published in *Prospects*, Vol. 11, no. 1).

Perry, W. (1976) *Open University: a Personal Account by the First Vice-Chancellor*, Open University Press, Milton Keynes.

Perry, W.G. (1970) *Forms of Intellectual and Ethical Development in the College Years: a Scheme*, Holt, Rinehart & Winston, New York.

Peters, O. (1973) *Die didaktische Struktur des Fernunterrichts. Untersuchungen zu einer industrialisierten Form des Lehrens und Lernens*, Beltz, Weinheim.

Peters, O. (1983) in D. Sewart, D. Keegan and B. Holmberg (eds.) *op. cit.*

Peters, O. (1989) The iceberg has not melted: further reflections on the concept of industrialisation and distance teaching, *Open Learning*, Vol. 4, no. 3, pp. 3–8.

Piaget, J. (1970) *The Science of Education*, Orion, New York.

Pillener, A. E. G. (1969) Multiple marking: Wiserman or Cox?, *British Journal of Educational Psychology*, Vol. 31, pp. 313–315.

Pinsonneault, A. and Kraemer, K. L. (1990) The effects of electronic meetings on group processes and outcomes: an assessment of the empirical research, *European Journal of Operational Research*, Vol. 46, pp. 143–161.

Postlethwaite, T. N. and Wiley, D. E. (1992) *The IEA Study of Science II: Science Achievement in Twenty Three Countries*, The International Association for the Evaluation of Educational Achievement, Pergamon, Oxford.

Purnell, R. (1993) Science for special needs, in E. Whitelegg, J. Thomas and S. Tresman (eds.) *op. cit.*

Radcliffe, J. (1990) Television and distance education in Europe: current roles and future trends, in A. W. Bates (ed.) *Media and Technology in European Distance Education*, EADTU, Open University, Milton Keynes.

Raggett, P. (1994) Outcomes and assessment in open and distance education, in F. Lockwood (ed.) *Materials Production in Open and Distance Learning*, Paul Chapman Publishing, London.

Raizen, S. (1993) Three decades of science reform in the USA, in D. Edwards, E. Scanlon and D. West (eds.) *Teaching, Learning and Assessment in Science Education*, Paul Chapman Publishing, London.

Ramsden, P. (1992) *Learning to Teach in Higher Education*, Routledge, London.

Reigeluth, C.M. (1969) Educational technology at the crossroads: new mind sets and new directions, *Educational Technology Research and Development*, Vol. 37, no. 1, pp. 67–80.

Reigeluth, C. M. (1983) *Instructional Design Theories and Models: an Overview of their Current Status*, Lawrence Erlbaum Associates, Hillsdale, NJ.

Reigeluth, C.M. (1993) New directions for educational technology, in T. O'Shea and E. Scanlon (eds.) *op. cit.*

Riley, J. (1979) I wonder what it's like to write a unit?, *Teaching at a Distance*, no. 14, pp. 1–8.

Riley, J. (1984a) The problems of drafting distance education materials, *British Journal of Educational Technology*, Vol. 15, no. 3, pp. 192–204.

Riley, J. (1984b) The problems of revising drafts of distance education materials, *British Journal of Educational Technology*, Vol. 15, no. 3, pp. 205–226.

Riley, J. (1984c) An explanation of drafting behaviours in the production of distance education materials, *British Journal of Educational Technology*, Vol. 15. no. 3, pp. 226–238.

Robinson, D. (1994) The virtual microscope, *Terra Nova*, Vol. 6, no. 6, pp. 638–641.

Robinson, B. (1994) in F. Lockwood (ed.) *Materials Production in Open and Distance Learning*, Paul Chapman Publishing, London.

Rodgers, M. (1992) The Hawking phenomenon, *Public Understanding of Science*, Vol. 1, no. 2, pp. 231–234.

Rogoff, B. and Lave, J. (1984) *Everyday Cognition: its Development in Social Context*, Harvard University Press, Cambridge, MA.

Romiszowski, A.J. (1981) *Designing Instructional Systems*, Kogan Page, London.

Romiszowski, A.J. (1984) *Producing Instructional Systems*, Kogan Page, London.

Romiszowski, A.J. (1986) *Developing Auto-Instructional Materials*, Kogan Page, London.

Romiszowski, A. (1988) *The Selection and Use of Instructional Media*, Kogan Page, London.

Ross, S.M. (1991) Interactive videodiscs for science education, *Journal of Computer Assisted Learning*, Vol. 7, no. 2, pp. 96–103.

Ross, S. and Scanlon, E. (1989) Conventional wisdom, course choice and student progress in science courses, *Open Learning*, Vol. 4, no. 3, pp. 14–22.

Ross, S. and Scanlon, E. (1995) Listening to the learner in physics, *Proceedings of the International Conference on Distance Education*, Birmingham, June.

Rowell, T. (1991) Women's examination results, *The Cambridge Review*, Vol. 100, pp. 94–95.

Rowntree, D. (1974) *Educational Technology in Curriculum Development*, Harper & Row, London.

Rowntree, D. (1977) *Assessing Students: How Shall we Know Them?* Harper & Row, London.

Rowntree, D. (1990) *Teaching Through Self Instruction* (2nd edn), Kogan Page, London.

Rumble, G. (1987) Why distance education can be cheaper than conventional education, *Distance Education*, Vol. 8, no. 1, pp. 72–94.

Rumble, G. (1989a) 'Open learning', 'distance learning' and the misuse of language, *Open Learning*, Vol. 4, no. 2, pp. 28–36. [Responses: Lewis, R. (1990) *Open Learning*, Vol. 5, no. 1, pp. 3–8 and Nation, D., Paine, N. and Richardson, M. (1990) *Open Learning*, Vol. 5, no. 2, pp. 40–46].

Rumble, G. (1989b) The role of distance education in national and international development: an overview, *Distance Education*, Vol. 10, no. 1, pp. 83–107.

Rumble, G (1989c) On-line costs: interactivity at a price, in R. Mason and A. Kaye (eds.) *Mindweave*, Pergamon, Oxford.

Rushby, N. J. (1979) *An Introduction to Educational Computing*, Croom Helm, London.

Russell, T. (1983) Analysing arguments in science classroom discourse: can teachers' questions distort scientific authority?, *Journal of Research in Science Teaching*, Vol. 20, pp. 27–45.

Säljö, R. (1989) Learning in educational settings: methods of inquiry, in T. Evans and D. Nation (eds.) *op. cit.*

Salomon, G. (1979) *Interaction of Media, Cognition and Learning*, Jossey-Bass, San Francisco.

Salomon, G. (1983) Using television as a unique teaching resource for OU courses, IET report on broadcasting, no. 225, internal document, Open University, Milton Keynes.

Sanders, N. (1966) *Classroom Questions*, Harper & Row, New York.

Sargant, N. (1988) Education and broadcasting, in N. Paine (ed.) *Open Learning in Transition: an Agenda for Action*, Kogan Page, London.

Sauve (1993) Media and distance education, in K. Harry *et al.* (eds.) *op. cit.*

Scanlon, E. (1981) Evaluating the effectiveness of distance learning: a case study, in F. Percival and H. Ellington (eds.) *Aspects of Educational Technology, XV*, Kogan Page, London.

Scanlon, E., Jones, A. and O'Shea, T. (1982a) Evaluating CAL at the British Open University, in A. Jones, E. Scanlon and T. O'Shea (eds.) *The Computer Revolution in Education*, Harvester Press, Lewes.

Scanlon, E., Jones, A., O'Shea, T., Murphy, P. and Whitelegg, E. (1982b) Computer Assisted Learning, *Institutional Research Review*, 1, Open University internal publication, Milton Keynes.

Scanlon, E. (1984) Preliminary organisers in a science course, in E. Henderson and M. B. Nathenson (eds.) *Independent Learning in Higher Education*, Educational Technology Publications, Englewood Cliffs, NJ.

Scanlon, E., Edwards, D. and West, D. (eds.) (1993) *Teaching, Learning and Assessment in Science Education*, Paul Chapman Publishing, London.

Scanlon, E., Edwards, D. and Whitelegg, E., with Ashby, A. and Brown, E. (1993b) in E. Whitelegg, J. Thomas and S. Tresman (eds.) *op. cit.*

Scanlon, E., O'Shea, T., Byard, M., Draper, S., Driver, R., Hennessy, S., Hartley, J. R., O'Malley, C., Mallen, C., Mohamed, G. and Twigger, D. (1993c) Promoting conceptual change in children learning mechanics, *Proceedings of the Nineteenth Annual Cognitive Science Conference*, Boulder, Colorado, June.

Scanlon, E. (1994) Study guide for Block 2, *ES821 Science Education*, Open University, Milton Keynes,

Scanlon, E., Murphy, P., Hodgson, B. and Whitelegg, E. (1994) A case study approach to studying collaboration in primary science classrooms, *Proceedings of the International Conference on Group and Collaborative Work*, Glasgow, September.

Schmid, R. F. and Telaro, G. (1990) Concept mapping as an instructional strategy for high school biology, *Journal of Educational Research*, Vol. 84, no. 2, pp. 78–85.

Schramm, W. (1977) *Big Media – Little Media*, Sage Publications, Beverley Hills, CA.

Scott, P. (1994) Recent developments in quality assessment in Great Britain, in D. Westerheidjen, J. Brennan and P. Maasen (eds.) *Changing Contexts of Quality Assessment: Recent Trends in West European Higher Education*, Lemma, Utrecht.

Screen, P. (1986) *Warwick Process Science*, Ashford Press, Southampton.

Scriven, M. (1967) *The Methodology of Evaluation*, American Educational Research Association Monograph Series on Curriculum Evaluation, no. 1, pp. 32–83.

Scriven, M. (1980) *Evaluation Thesaurus* (2nd edn), Edgepress, Inverness.

Self, J. (1985) *Microcomputers in Education: a Critical Appraisal of Educational Software*, Harvester Press, Brighton.

Self, J. (1987) *Microcomputers in Education: a Critical Appraisal of Educational Software* (2nd edn), Harvester Press, Sussex.

Sellen, A. (1992) Presentation to the British Association for the Advancement of Science annual meeting, Southampton, September.

Sewart, D. (1983) Distance teaching: a contradiction in terms? in D. Sewart, D. Keegan and B. Holmberg (eds.) *op. cit.* (reprint of 1981 article originally published in *Teaching at a Distance*, no. 19, pp. 8–18).

Sewart, D., Keegan, D. and Holmberg, B. (eds.) (1983) *Distance Education: International Perspectives*, Croom Helm, London.

Sharples, M. (1987) The design of a user friendly system, in A. Jones, E. Scanlon and T. O'Shea (eds.) *The Computer Revolution in Education*, Harvester Press, Brighton.

Shayer, M. and Adey, P.S. (1981) *Towards a Science of Science Teaching*, Heinemann Educational, London.

Shulman, L. S. and Keisler, E. R. (1966) *Learning by Discovery*, Rand McNally & Co., Chicago.

Slavin, R. (1983) *Cooperative Learning*, Longman, New York.

Smith, P. (1987) Distance education and educational change, in P. Smith and M. Kelly, *op. cit.*

Smith, P. and Kelly, M. (1987) *Distance Education and the Mainstream*, Croom Helm, London.

Smith, R. (1986) The alternate reality kit: tensions between literalism and magic, in O. Boyd-Barrett and E. Scanlon (eds.) *Computers and Learning*, Addison Wesley, Wokingham.

Smith, R., O'Shea, T., O'Malley, C., Scanlon, E. and Taylor, J. (1991) Preliminary experiments with a distributed, multi-media problem-solving environment, in J.M. Blowers and S.D. Benford (eds.) *Studies in Computer Supported Cooperative Work: Theory, Practice and Design*, Elsevier, Oxford.

Smith, R. (1993) A prototype futuristic technology for distance education, in T. O'Shea and E. Scanlon (eds.) *New Directions in Educational Technology*, Springer Verlag, Berlin.

Solomon, J. (1989) in P. Adey, J. Bliss, J. Head and M. Shayer (eds.) *Adolescent Development and School Science*, Falmer Press, Lewes.

Spensley, F., O'Shea, T., Singer R., Hennessy, S., O'Malley, C. and Scanlon, E. (1990) A direct manipulation microworld for sliding friction, CITE Report no.105, Open University internal report, Milton Keynes.

Stake, R. E. (1967) The countenance of educational evaluation, *Teachers' College Record*, no. 68, pp. 523–540.

Stake, R. E. (1978) The case study in social inquiry, *Educational Researcher*, Vol. 7, no. 2, pp. 5–7.

Stanford, J. (1980) A one-person course team, *Teaching at a Distance*, no. 18, pp. 3–9.

Starfield, A. M., Smith, K. A. and Blelach, A. L. (1990) *How to Model it*, McGraw Hill, New York.

Stoane, C. and Stoane, J. S. (1981) Answers at a distance, in F. Percival and H. Ellington (eds.) *Aspects of Educational Technology XV: Distance Learning and Evaluation*, Kogan Page, London.

Stokes, M. J. and Stafford, K. J. (1986) Teaching effectiveness in the science laboratory, *Proceedings of the Hong Kong Educational Research Association Annual Conference*, Hong Kong.

Stratton, N., Armitage, P. and Scott, B. (1982) Reports of the Student Assessment Research Group, IET internal document, Open University, Milton Keynes.

Stringer, M. (1980) Lifting the course team curse, *Teaching at a Distance*, Vol. 18, pp. 13–16.

Stufflebeam, D. (1971) *Educational Evaluation and Decision Making*, F. E. Peacock Publishers, Itasca, IL.

Sumner, H. (1991) BBC education: the effectiveness and evaluation of educational broadcasting, *Journal of Educational Television*, Vol. 17, no. 1, pp. 159–171.

Svensson, L. (1977) On qualitative differences in learning: III. Study skill and learning, *British Journal of Educational Psychology*, Vol. 47, pp. 233–243.

Tan, C. (1992) An evaluation of the use of continuous assessment in the teaching of physiology, *Higher Education*, Vol. 23, pp. 255–272.

Taylor, E. (1984) Overview: developing study skill, in E. S. Henderson and M. B. Nathenson (eds.) *op. cit.*

Taylor, J. (1991) Using television in distance teaching, Audio-Visual Media Research Group, AV pack no. 4, Open University internal publication, Milton Keynes [revision of Bates (1985) *op. cit.*].

Taylor, J. (1992) Drop-in viewing audience survey, Programme on Learner Use of Media report no. 40, Open University internal publication, Milton Keynes.

Temple, H. (1988) in N. Paine (ed.) *op. cit.*

Teshome, A. (1994) Exploring the potential of expert systems in distance education: the case of Wye College external programme student support service, unpublished PhD thesis, University of London.

The Royal Society (1985) Science *is* for everybody. Executive summary from *The Public Understanding of Science*, The Royal Society, London.

Thorndike, E. (1913) *Educational Psychology, Vol. 11: the Psychology of Learning*, Teachers College, Columbia, New York.

Thorpe, M. (1994) Planning for learner support and the facilitator role, in F. Lockwood (ed.) *Materials Production in Open and Distance Learning*, Paul Chapman Publishing, London.

Toothacker, W.S. (1983) A critical look at introductory laboratory instruction, *American Journal of Physics*, Vol. 51, no. 6, pp. 516–520.

Trow, M. (1972) *The Expansion and Transformation of Higher Education*, General Learning Press.

Twigger, D., Draper, S., Driver, R., Hartley, R., Hennessy, S., O'Malley, C., O'Shea, T. and Scanlon, E. (1991) The conceptual change in science project, *Journal of Computer Assisted Learning*, Vol. 7, no. 2, pp. 144–155.

Tyler, R. (1934) *Constructing Achievement Tests*, Ohio State University, Columbus.

Tyler, R. (1986) Another paradigm and a rationale, in *Changing Concepts of Educational Evaluation*, International Journal of Educational Research Monograph, Vol. 10, p. 1.

UNESCO (1983) *Science for All*, UNESCO Regional Office for Education in Asia and the Pacific, Bangkok.

Van Patten, J., Chao, C. I. and Reigeluth, C. M. (1986) A review of strategies for sequencing and synthesizing instruction, *Review of Educational Research*, Vol. 56, no. 4, pp. 437–471.

Veasey D'Souza, P. (1991) The use of electronic mail as an instructional aid: an explanatory story, *Journal of Computer-Based Interaction*, Vol. 8, no. 3, pp. 106–110.

Viennot, L. (1979) Spontaneous reasoning in elementary dynamics, *European Journal of Science Education*, Vol. 1, no. 2, pp. 205–221.

Vygotsky, L. S. (1978) *Mind in Society; the Development of Higher Psychological Processes*, Harvard University Press, Cambridge, MA.

Waller, R. (1979) Typographic access structures for educational texts, in J. Kohlers *et al.* (eds.) *Processing Visible Language*, Vol. 1, Plenum Press, New York.

Waller, R. and Lefrere, P. (1981) New technologies for academic publishing, *Teaching at a Distance*, Vol. 19, pp. 32–39.

Waller, R. (1984) A structured exercise in biology – criticism and alternatives, in E. Henderson and M.B. Nathenson (eds.) *Independent Learning in Higher Education*, Educational Technology Publications, Englowood Cliffs, NJ.

Waterman, D. and Thompson, S. (1995) *Smallscale Chemistry*, Addison Wesley, Wokingham.

Watts, M. (1991) *The Science of Problem-Solving*, Cassell, London.

Wedemeyer, C. (1977) Independent study, in A. S. Knowles (ed.) *The International Encyclopaedia of Higher Education*, pp. 548–557, CIHED, Boston.

Weir, S. (1987) *Cultivating Minds*, Harper & Row, London.

Wellington, J. (ed.) (1989) *Skills and Processes in Science Education*, Routledge, London and New York.

Wells, R. (1992) *Computer-Mediated Communication for Distance Education: an International Review of Design, Teaching and Institutional Issues*, Research Monograph no. 6, American Center for the Study of Distance Education, Pennsylvania State University.

Wenger, E. (1987) *Artificial Intelligence and Tutoring Systems*, Morgan Kauffman, Los Altos, CA.

Weston, D. (1993) Shared whiteboards, presentation at the CAL 2000 meeting, Open University, Milton Keynes.

Whitelegg, E., Thomas, J. and Tresman, S. (eds.) (1993) *Challenges and Opportunities for Science Education*, Paul Chapman Publishing, London.

Wild, R. and Winniford, M. (1993) Remote collaboration among students using electronic mail, *Computers and Education*, Vol. 21, no. 3, pp. 193–203.

Williams, M., Holland, J. and Stevens, A. (1981) An overview of STEAMER: an advanced computer propulsion system for propulsion engineering, *Behaviour Research Methods and Instrumentation*, Vol. 13, No. 2, pp. 85–90.

Willis, B. (ed.) (1994) *Distance Education: Strategies and Tools*, Educational Technology Publications, Englewood Cliffs, NJ.

Wilson, H. (1963) Speech to a Labour Party rally, Glasgow, 8 Sept 1963, quoted in Perry, H. (1976) *op. cit.*

Winn, W. (1993) A constructivist critique of the assumptions of instructional design, in T. Duffy, J. Lowyck and D. Jonassen (eds.) *Designing Environments for Constructive Learning*, Springer Verlag, New York.

Woolnough, B. and Allsop, T. (1985) *Practical Work in Science*, Cambridge University Press, Cambridge.

Woolnough, B. E. (1989) Towards a holistic view of processes in science education, in J. Wellington (ed.) *op. cit.*

Woolnough, B. (ed.) (1991) *Practical Science*, Open University Press, Milton Keynes.

Yankelovich, N., Meyerowitz, N. and Van Dam, A. (1985) Reading and writing the electronic book, *Computer*, Vol. 18, no. 10, pp. 15–29; reprinted in O. Boyd-Barrett and E. Scanlon (eds.) (1990) *Computers and Learning*, Addison Wesley, Wokingham.

Ziman, J. (1986) Science education for whom? in J. E. Brown, A. Cooper, T. Horton, F. Toates and D. Zeldin (eds.) *Science in Schools*, Open University Press, Buckingham.

Zimmer, R. (1984) The relational glossary – an aid to learning and review, in E. S. Henderson and M. B. Nathenson (eds.) (1984) *Independent Learning in Higher Education*, Educational Technology Publications, Englewood Cliffs, NJ.

INDEX